BICTON COLLEGE OF AGRICULTURE
BICTON
LIBRARY
14418
344.41
05270
FOR
REFERENCE ONLY
WITHDRAWN
BICTON
COLLEGE
of Agriculture
CULTUS EX MENTIS CULTU
Education Services
Centre
KT-173-486

The COSHH Regulations: A Practical Guide

The COSHH Regulations: A Practical Guide

Edited by

Diana Simpson, MPhil, PhD, FIFST, CEng FPRI, FRSH, CChem FRSC

W. Gordon Simpson, MA, MPhil, AMSIA, FIDiagE, FPRI

A catalogue record for this book is available from the British Library.
ISBN: 0-85186-189-X

Published by The Royal Society of Chemistry,
Thomas Graham House, Science Park, Cambridge CB4 4WF

Printed by Redwood Press Ltd, Melksham, Wiltshire

Preface

Concern for the health and well-being of employees is no new thing and there is a long and honourable tradition in British industry of well-lit and well-ventilated factories, with associated welfare provision such as canteens, medical centres, and sports facilities. Some of the best employers were pioneers in these matters.

However, standards vary and are not always good. If a business is highly-profitable it can afford to provide better buildings and generally better facilities than others less well-run or less fortunate. Accidents occur—sometimes perhaps as a result of negligence—and life in general becomes more complicated. Work in all environments, not only factories but also on farms and in offices, involves an increased use of machinery and of manufactured products. In the past an operator would know that he was handling cotton or woollen fibre—or at least the general nature of the material being worked—but often (if not usually) today, employees may well have little knowledge of the underlying nature of the goods they deal with—such as the composition of a plastic, a rubber, a synthetic textile, or of a coating on paper—and this absence of knowledge can lead in time to concern as to the 'unknown' effects of these substances. It happens too that materials thought at one time to be inert or 'safe' are found later to be harmful in some way. Hence there has been growing need, real as well as imagined, for scientific investigation of working environments, and for factual assessments of what is or might be found there.

Possibly also, as many maintain, there has been a need for more elaborate and comprehensive legislation—which Parliament in Britain and the European Commission at Brussels (among others) have been endeavouring to supply.

This book presents a review derived from our own experiences as scientific consultants active in the field of health and safety for many years, together with contributions by other experienced authors on topics such as the development in recent times of legislation in this field in Britain, and in particular of the requirements and knowledge gained so far of the application of Regulations for 'Control of Substances Hazardous to Health' (known, it seems, almost universally as 'COSHH').

It should be explained at this stage that (in common with many other consulting practices) whenever our practice undertakes any kind of work it does so on a confidential basis. We take our relationship with the client to be professional and similar to that of a medical practitioner or even of a spiritual adviser; all members of our staff sign agreements not to reveal even the names of clients, and samples in the laboratory are identified only by work numbers; if necessary we would feel obliged to resist in the courts any compulsion to open our records to a third party. This of course is not to say that our recommendation to a client might not be to report an incident or a finding to the proper authority—but it is felt to be primarily the responsibility of the client rather than ours to do this. We make clear to the clients that employees of HM Customs & Excise are empowered to see copies of our invoices (and in certain circumstances could force a way into the laboratory for this purpose) but apart from official action such as this a client's contacts with us are as secure as is possible for a non-government agency. (We feel sometimes that independent experts should have comparable rights to search for documents in official hands, or at least to have them brought before a court, but the balance of power at present is somewhat the other way.)

Because of the wish to keep confidence it is necessary for us to disguise to some extent the information given here in the examples of work done. However, the changes do not affect the character and the validity of the results. When associated circumstances are mentioned they are not specific and similar ones have been encountered quite frequently.

The volume is meant to be helpful (and, indeed, to be 'a practical guide') but we must make clear at the outset that although it offers information and advice in good faith no responsibility for these or for their application is accepted. No guarantee is given nor should be taken to be implied—by the Royal Society of Chemistry, by the editors, or by any individual author.

For any readers inclined to see sexual chauvinism everywhere might we just add that on occasions when we write 'he' we mean *always* 'he or she'; it is just that some sentences would be too cumbersome if we tried to make this distinction every time.

We have been fortunate in the selection of contributors. Very few who agreed to contribute failed to do so and we feel that all the work recorded here reflects first-hand knowledge and experience. The contributors (not to mention their secretaries) have been forbearing with our queries and helpful with the editorial work—and we are very grateful to them.

In Chapter 2, John Banham describes how the Regulations were brought into being and gives them a cautious welcome, although expressing concern over the possible cost to industry.

S. G. Luxon presents in Chapter 3 a comprehensive review of the legal requirements; this is followed by an examination by Dr J. G. Firth of the special needs for measurement arising from the Regulations and from the Approved Codes of Practice.

The requirements for records provides the topic for two chapters—Dr P. J. Hewitt on 'Statutory Records' (Chapter 5), and Paul Hopkin on the recording of assessments (Chapter 6).

In Chapter 7, Dr C. H. Collins, MBE, reviews the microbiological aspects of applying the Regulations in practice, and all but one of the subsequent chapters cover practical experience in a variety of fields—newspaper printing (Chapter 8) by S. King, retail business (Chapter 9) compiled by the editors, and a research environment (Chapter 10) by J. P. Samuels.

Finally (Chapter 11), R. W. Hazell and co-authors relate the requirements of the COSHH assessment to the larger question of maintaining the quality of the environment at large.

D. Simpson
W. G. Simpson

Contents

Contributors

John Banham
Director-General, Confederation of British Industry
C. H. Collins, MBE, DSc, FRCPath, CBiol FIBiol
Consultant
J. G. Firth, BSc, PhD, CChem FRSC
Director, The Occupational Medicine and Hygiene Laboratories, Health & Safety Executive
R. W. Hazell, BSc, CChem MRSC
Royal Society of Chemistry
P. J. Hewitt, PhD, MSc, ARTCS, MInstRP, FIOSH, FIOH, CPhys MInstP, CChem FRSC
Principal Occupational Health and Safety Consultant, Bradford University Research Limited
P. Hopkin, BSc, FIOSH, RSP, FIRM, CertEd
Principal Consultant, Progressive Consultants
S. King, DipSM, MRSH, MIOSH, MIIRSM, FIPC
Personnel Health & Safety Consultants Limited
S. G. Luxon, FRSH, CChem FRSC
Consultant in health and safety
M. L. Richardson, BSc, CChem FRSC
Birch Assessment Service for Information on Chemicals (BASIC)
J. P. Samuels
Safety Officer, BP Research

CHAPTER 1

Introduction

The Editors

Investigation of Working Environments

In Chapters 12 and 13 of an earlier book—'An Introduction to Applications of Light Microscopy in Analysis'—we referred at some length to investigations of working environments under the Health and Safety at Work Act, 1974, and to the principles we felt should underlie effective presentation of evidence in matters such as these.

Rather than repeat here the same information we refer readers to the earlier volume. However, that book was written before the COSHH Regulations came into being at a time when no one had experience of applying them in a practical sense. We take the opportunity here to record some more recent experiences.

Physiology and Psychology

It is said often that people 'work to live', not the other way round, but working is a habit very deeply ingrained and some people at least are not happy unless engaged in it—either for reward as employees or suppliers of services, or perhaps just to be active in some way.

The act of working implies a certain amount of wear and strain on the system—regardless of whether one is considering hard physical labour or, say, modern 'keyboard skills'—affecting muscular development, physique, the nervous system, and the brain. It takes time; one's private life, rest and recreation must be fitted into the hours left over after work. These hours may be crowded, especially if (as often is the case nowadays) the place of work is at a distance from home and more minutes, or hours, have to be given over each day to travelling. It is possible that employment, measured in terms of output per person, is more productive today than ever in the past, but this progress is accompanied by greater demands on the staff concerned, arising both from immediate conditions in which they work and also from pressures on their lives in general.

Happily, some seem to thrive under demands and pressures, but others do not. From the point of view of the practitioner in health and safety there may be difficulty in separating physiological problems which arise in 'normal' situations (that is, under conditions which present no difficulty for others) and those which may be attributed with confidence to circumstances at the place of work.

Psychology may complicate factual assessment in a variety of ways (as an example, forms of 'mass hysteria' have been noted in certain environments—such as where several inexperienced young people were working closely together) and a few unfortunate individuals, otherwise normal, can exhibit or develop unusual sensitivity to particular substances and situations.

Self-Sacrifice

It is a quality, and perhaps also a danger, that certain characters stand out in a group. Bernard Shaw said often that he did not wish to die (or, as he put it, 'to be thrown on the scrap heap') until he was thoroughly used up. The vastness of his literary output is an indication not only of a long life but also of a well-developed propensity to work. This was an exceptional person but there are those who are prepared to sacrifice themselves for the satisfaction of building a business or of defending an Empire—or even (on another level) who will risk their health for reasons which seem to others not so important, such as by climbing into a machine to replace a blocked filter. It may be that the Act of 1974 placed a duty on such people not to behave like this, and certainly it placed a duty on employers to restrain (so far as they can) such flamboyant conduct. The employer must try to prevent the exposure of employees—and of others, including members of the public at large—to the risks involved and the consequences of over-enthusiasm.

The Alleged Heartlessness and Rapacity of Employers

Staff may be keen, loyal, and conscientiousness—or (on the other hand) may be induced by bonus rates to cut corners and so raise output; sometimes there will be extra pressure to complete a task in time (for reasons such as to raise the number of deliveries when a financial period draws to a close, to complete contracts by dates agreed, or to load cargoes before ships sail). All such pressures call for careful consideration and explanation. Employers often are perceived as hard and heartless, and of course they are in a better position than many staff to appreciate what is necessary and to take account of possible consequences of their actions.

Our own experience suggests employers in general not to be heartless, though perhaps at times a little thoughtless. Perhaps some should be more alert to dangers. In the past, Marxism had a powerful influence on Western political ideas—and it asserts, among other things, that employers take for themselves 'surplus value' created by labour—in other words, that the employer is an exploiter and natural enemy of the employee. The notion frequently is reflected in debate and in legislation, with references to 'the two sides of industry', 'them and us', and so forth. There may be a serious political dimension associated with an industrial dispute, and often there will be a political undertone to which management and the scientist assisting should be sensitive.

Loyalty and Foolhardiness of Employees

A certain degree of pressure is associated with work and with any responsibilities undertaken knowingly—not only, say, to achieve a required level of output but also if necessary to tighten a specification, to modify a product, or to repair and set up a machine on time. Such matters are part of daily life and need to be dealt with smoothly and without harm to those concerned. Underlying this is the implication that methods and systems in use at the unit must be of capacity sufficient and condition good enough to meet extra demands on them.

By now it should be reasonable to think that the average employee is more aware than in the past of the need to be careful; from the point of view of the responsible manager the essence is to restrain the bravery and, indeed, foolhardiness that may be displayed sometimes and to lay emphasis on acting always in a methodical and practical manner.

Unseen Responsibilities

Managers carry also responsibilities of which they may be largely unaware. One of the first prosecutions under the Regulations involved effects from oxides of carbon from an engine used on a temporary basis (normally the substances would not have been present). A story was told some time ago of an able engineer with an important appointment which required the maintenance and safety of many bridges, water towers, and other installations. One of his towers failed, giving rise to damage, and because of this he was dismissed; probably he had not even seen the unit concerned (it would have been inspected by a junior) but he was responsible for it.

Changing Ideas in Occupational Hygiene

Occupational hygiene has been a pre-occupation of industrialists and reformers for longer than many of us realize, and it is useful to recall a little of this background. In Britain in the last century, when land and building materials were cheap and business usually was good, the approach often was to build for industrial use tall, airy structures, well-lit and well-ventilated.

Windows of plain glass would occupy much of the walls so there would be ample natural daylight; ceilings would be high, there was plenty of roof space, and the windows could be opened to the outside air. Owners would take pride in the design, appearance, and quality of their properties. Substantial materials were used in construction—frequently brick, slate, stone, heavy timber, and cast iron.

Today, when building is more costly, interior volumes are limited so far as practicable, windows may be omitted (so that inside only artificial light will be available), and frequently natural ventilation is replaced by some form of air re-circulation—to keep down the costs of heating. The range of materials used is more diverse than in the past; often they are less substantial in character than formerly—lighter in weight, or perhaps comprising fabrications from paper, plastic films, sheets, boards, and composite products. A typical industrial structure might comprise essentially a steel frame, walls of so-called 'in-fill' panels, and a sheet roof. Synthetic materials have been used increasingly for construction, both as elements of buildings (especially as panels, doors, furnishings, and insulation), and also for decorative effects (for ceilings, paints, and other finishes).

More complex and sophisticated compositions have become commonplace too in manufacturing units, training establishments, retail shops, and other places of business. It is worth remembering that a wide range of substances may be encountered not only within buildings but out-of-doors, in ships, aircraft, all manner of vehicles, and in other structures and objects. A Victorian artisan might come into contact with a range of common natural compounds—such as cellulose fibres, coal dust, and their decomposition products—but the range today is much wider and more complex.

Parallel with the use of a greater variety of substances, mixtures, and compounds, has been a great extension of interest in, concern with and detailed understanding of toxicology—which in turn has led in some instances to prohibitions and in others to increasingly strict controls over the conditions and ways in which particular substances may be handled and applied.

A hundred years or more ago the typical factory or office would be far more crowded with people than is the case today, when increased costs of labour and the effects of mechanization have combined to reduce drastically the numbers of operators engaged for a given level of output. In those days, not only were techniques for establishing and maintaining standards of occupational hygiene less sophisticated than now but there was a different emphasis—on screening large numbers of people as quickly as possible rather than more specific, individual consideration. Further, nineteenth century commerce and industry in Britain employed young people and children (as they are employed still overseas), many children starting to work half-time on reaching the age of ten years, whereas the current 'occupational exposure limits' are taken to be applicable for adults.

The Regulations

Parliament always has interested itself in the conditions under which British people were (and are) required to work, in factories and in other environments; as with most (if not all) activities of government the apparatus of control in this field has become more elaborate as time went on.

The Regulations for 'Control of Substances Hazardous to Health' were made under powers conferred by the Health and Safety at Work Act 1974. They were in preparation for some years and in a statement the Health and Safety Executive said it regarded them as the most important legislation in the field since the Act. They would provide a broad framework, replacing specific regulations covering specific substances and places of work. They were intended to be 'flexible' (in case changes in standards of control should be thought necessary in future), and to allow the repeal of certain legislation not in keeping with modern needs. They would allow government to meet the terms of the European Directive on protection of workers against risks from exposure to chemical, physical, and biological agents, and to ratify the International Labour Organization Convention 139 on carcinogens. The Regulations would be supported in due course by a number of codes of practice—covering, among other things, carcinogens, fumigation, and vinyl chloride.

Statutory Instrument 1988 No. 1657 'The Control of Substances Hazardous to Health Regulations 1988' was signed by the Parliamentary Under Secretary of State at the Department of Employment, Patrick Nicholls, on 26th September 1988 and laid before Parliament on 12th October 1988. The Regulations were intended to come into force in 1st October 1989.

The 'Explanatory Note' reads, on part:

> The terms used in the Regulations, including the term 'substance hazardous to health' are defined in regulation 2. By regulation 3 duties are imposed on employers (who for the purposes of these Regulations include self-employed persons) for the protection of their employees and of other persons who may be affected by their work.

However, as a matter of fact (and on the contrary), Regulation 2 does not define 'substance hazardous to health' (it defines neither 'substance' nor 'hazard'). It is of interest to note that this section is called not 'Definitions' but 'Interpretation'; it repeats the phrase used throughout the document—'substance hazardous to health'—followed by five classifications, some specific and some not at all specific, such as:

(*c*) a micro-organism which creates a hazard to the health of any person

(*d*) dust of any kind, when present at a substantial concentration in air

(*e*) a substance, not being a substance mentioned in sub-paragraphs (*a*) to (*d*) above, which creates a hazard to the health of any person which is comparable with the hazards created by substances mentioned in those sub-paragraphs.

(It is written in another place that 'substance' means 'any natural or artificial substance whether in solid or liquid form or in the form of gas or vapour

TABLE I

Term	*Meaning*
Substance or substances hazardous to health	Anything that the Health and Safety Executive may decide
Pollution	Anything the Pollution Inspectorate may decide
Nuisance	Anything else that can be prosecuted
Poisons	A limited range of toxic substances inconveniently defined in law
Toxic substance	Anything anyone happens to dislike

(including micro-organisms).') In other words, a substance is a substance; a hazard is a hazard. (This must be what was meant by 'flexibility'.)*

It should be clear already that the Regulations confer ill-defined but very broad powers. Apart from their not saying precisely what is meant by 'substance hazardous to health' (presumably it must mean *any* substance) there are questions such as interpretation of 'a micro-organism', 'any person', 'substantial concentration', and a hazard 'comparable with the hazards created by substances mentioned'. At one time it was thought that one of the benefits of advances in scientific measurement would be specific legislation, in which limits could be stated numerically and verified in practice, but with these Regulations this does not appear to be the case.

In a somewhat sardonic comment, an ephemeral publication for consulting scientists gave the five relevant 'definitions' shown in Table I.

Requirements

The Regulations require that for all work being carried on before the end of 1989 an assessment of the risks to health of the employees concerned must be made before 1st January 1990. The assessment must be 'suitable and sufficient' and must include consideration of 'steps that need to be taken to meet the requirements of these Regulations'.

Various measures are cited for the 'control of exposure to substances hazardous to health'. The object should be to 'prevent' exposure to substances but it is conceded that when this is not possible 'adequate control' should be provided. Specific indications are given as to when 'control' will not be regarded as 'adequate'.

'All reasonable steps' must be taken to ensure that the control measures or protective equipment are used and applied properly. Checks must be made of equipment (at intervals stated) and there must be monitoring of the workplace. 'Suitable procedures' must be used and records kept; when records refer to 'the personal exposures of identifiable employees' they must

*The 'Concise Oxford' was hardly more helpful: among several definitions of 'substance' perhaps the most appropriate was '*4*. Particular kind of matter'; 'hazard' was 'Game at dice', 'Expose to hazard' was 'venture on', and 'hazardous' was 'Risky; dependent on chance'.

be kept for at least 30 years. Employees must be kept under 'suitable health surveillance'. Information about exposure to 'substances hazardous to health' must be provided to employees, including information about risks and precautions to be taken. There are special provisions with regard to fumigation, and exemptions for certain types of work.

Penalties

There is no reference in the Regulations to penalties but they are as in the Health and Safety at Work Act and may be summarized briefly:

(*a*) on summary conviction (in a magistrate's court)—a maximum fine of £2000 on each charge

(*b*) on conviction on indictment (if applicable)—fines of unlimited amount and a term of imprisonment not exceeding two years on each count.

At the time of writing, there have been prosecutions after two men were overcome by carbon monoxide given off by a portable petrol-engined generator, and of a company which was using caustic soda to clean graffiti from a war memorial. In both instances pleas of 'guilty' were entered; fines of £200 and £1000, respectively, were levied.

Concerns Expressed

Early in 1990 several peers commented adversely in the course of a debate in the House of Lords on the great volume of legislation that was being brought forward, and (in some instances at least) on the indifferent quality of the drafting. Lord Rippon of Hexham, a former minister, noted that in 1989 there were 2581 pages of statutes and in 1988 2170 pages; these figures might be compared, he said, with 790 pages in 1977 and 830 pages for 1978. He commented: 'The sheer volume and complexity of legislation today is such that it is quite impossible for either House to scrutinize or to discuss it properly. It constitutes a new form of despotism, giving increasing power to the Executive.' Other speakers suggested that it was undesirable if government enacted general legislation and went on to fill in details later: bad legislation could lead to chaos and to a need for successive amendments.

SI 1657 was a tiny part only of the legislative stream (as a matter of fact, 23 pages) but it is fair to write that it caused some concern, especially perhaps among employers and managers with technical knowledge enough to appreciate all that might be involved—as examples, some manufacturing chemists, educational, and research establishments. It has been noted already that the Regulations were several years in preparation, a period in which there was discussion and argument over the texts of a series of drafts. However, when signed eventually they were little changed in effect and appeared to encompass any and every substance.

An item in *Chemistry in Britain*, December 1989, reflected the concern and summarized questions asked and replies given at meetings about the Regulations which were arranged in various parts of the country for members of the

Royal Society of Chemistry. Comments made at these meetings included doubts as to the competence of some staff to make the assessments, coupled with suggestions by representatives of the Society for an approach which might be adopted in a laboratory or other establishment in which very large numbers of substances were being held in stock, even in small or tiny amounts (for example as standards for purposes of analysis). In cases such as these it was suggested that an assessor might try classifying the substances according to the extent to which they could represent hazards, then to assess the groups or classifications as entities rather than substance by substance. It was suggested that new compounds or other materials about which no information was available could be regarded as falling into the most hazardous group.

It seemed from the questions reported by *Chemistry in Britain* that not everyone understood that the degree of toxicity or 'hazard' associated with a substance was not the subject for assessment—it was the extent of the hazard under the conditions of use at the establishment concerned. Thus the same substances could be assessed quite differently in different circumstances, even within establishments of the same organization.

There were numerous requests for a national data bank on toxicology but it was pointed out in reply that there were already many sources of information about toxicology and while a single source might be desirable it was not the aspect of health and safety on which emphasis should be placed in order to satisfy these Regulations.

Many of the participants in the meetings asked advice on what might be considered 'suitable and sufficient'—but it was said that it was a question that could not be answered as yet. Ultimately, guidance would be provided by rulings in the courts; meanwhile, the onus was on the employer or manager to cover the business and himself so far as possible. Some assistance might be given by relevant sections of the codes of practice and other advisory publications and meanwhile it might be wise to regard the assessment as a form of insurance against possible difficulty in the future.

Quite a number of further questions come to mind (some of them might well have been asked but were not reported in the summary published), including:

(i) how can I assess the hazard (if any) associated with vapours or products of decomposition under the conditions of processing at my establishment?

(ii) how might I obtain detailed information about changes in conditions of processing; do they vary in the course of a working day, in Winter or Summer, or during the night shift? What effect do such changes have on the substances concerned?

(iii) how might I assess the possibility of hazard arising from reactions between two or more of the substances used in my unit, including reactions between substances which would only come into contact by accident?

(iv) how do I assess the risks associated with the use of mixtures of substances, especially if the information I have is for the substances alone?

(v) can something regarded as 'inert'—such as a filler or bulking agent in a pharmaceutical or detergent product—or even harmless, represent a hazard in some circumstances?

(vi) substances such as metals and plastics are processed at high temperatures and often also under pressure; for example, when hot steel is spilt or poured on water or even on damp concrete a violent explosion can ensue.

(vii) how do I assess hazards for materials of which I am unaware—such as cleaning compounds brought by employees on their own initiative, or substances used by contractors—and for which we have no information?

(viii) some substances enter my place of business fortuitously—as examples, pollen from parks or window-boxes, samples or other items brought in by visitors, exhibition material, wind- or water-borne substances and organisms; what is to be done about these, and to what extent could I be expected to anticipate them?

A correspondent in *The Spectator*, 22.9.90. (Letters, page 32), anticipated problems to be experienced by nurserymen and garden centres—and by employers in any field who tried to make their facilities 'green'—observing that many garden plants were poisonous:

> . . . cyanides, alkaloids, and deadly peptides and terpenoids are all produced in the English town or country garden.
>
> A principle of COSHH is that the accused employer is presumed guilty unless he can prove himself innocent (which he is unlikely to do). Employers of jobbing gardeners, or those who beautify approaches to factory or office with gardens, seem to me to be left in an invidious position.
>
> About half the plants in my own employer's gardens are poisonous, others are allergens to the susceptible; the only solution to his responsibilities under COSHH appears to be (chromate-free) concrete.

There is confusion again here between the toxicology of the substances concerned (plant materials) and the hazards they present in practice under the conditions of work: on the other hand, it could well prove unwise not to take due notice of the point made.

Quality of Management

Much is heard of the 'quality of management' and the alleged strengths or (more usually) the short-comings of the managers themselves. No doubt a responsibility for health and safety is one of the more important aspects of good management (and nowadays an accident even might be held to indicate short-comings in this regard).

Nevertheless it does seem a little hard to heap upon unfortunate individuals responsibilities which in fact cannot be discharged. The Regulations take it that management will be aware of every substance present in an establishment but this raises certain difficulties which we summarize thus:

(i) purchasing may not be centralized, and so there may be no comprehensive list of every type of material bought by a unit

(ii) it will not be sufficient to require managers to notify every change in buying policy (they may switch between suppliers quite frequently, even from day to day in some cases); either central records will be needed for all consignments purchased, regardless of source, or limitations will have to be placed on the buyers so as to avoid changes which have not been authorized in advance

(iii) frequently, methods of manufacture and the formulations of products are changed without notice, and without informing buyers; thus, a product which ostensibly is the same as in the past is altered without users knowing it. It would be desirable if possible to prevent suppliers from doing this, or to require them at least to advise the unit of any changes in composition

(iv) an establishment using natural products (as distinct from manufactures) must anticipate frequent and quite wide variations in their composition; also, such natural products may contain both synthetic and natural toxins, together with micro-organisms and perhaps other hazardous items at random (like tarantula spiders in consignments of bananas)

(v) it will be necessary to prevent all employees, contract labour, salesmen and other visitors from bringing any substances to the premises except by prior agreement

(vi) possibly even regular screening of staff, live- and dead-stock for micro-organisms might be necessary.

Even assuming a management with sufficient strength of purpose to meet all such eventualities we suspect there would remain still a likelihood of something unexpected actually happening.

In this connection, we recall an incident in which a large national organization did not specify limits for residues in a synthetic material on which there were two or more organic coatings (the coatings themselves were specified, but not the residues in them—that is, the maxima for substances which might be left in or on the material after it was made and coated). No one anticipated that such residues might reach harmful levels, but they did.

Certainly, in light of the Regulations, an organization is well-advised not to buy anything that is not specified in detail in writing.

Anticipating the Worst

It seems reasonable to suppose that in some degree at least the Regulations will be applied capriciously—as, indeed, other legislation is applied (observation suggests, for example, that many more motorists exceed speed limits than actually are prosecuted in the courts). In a practical sense this implies that employers and managers need to take care not only to satisfy the requirements but to be able to show, if necessary, that this was done. An obvious element in so doing is maintaining a close and amicable relationship with the Health and Safety Inspectorate and with all other relevant officials, obtaining in writing their comments and approval of action taken with regard to safe working. A business of any size and at any level should remember that the local offices of the Health and Safety Executive, HM Customs and Excise, HM Pollution Inspectorate, HM Petroleum Inspectorate, and the local authorities all are accumulating gradually files of information covering each organization in the area. It is important from the point of view of the businessman or manager that those files should be as reliable and bland as possible. They will include records of visits for inspection, which usually are unannounced and unexpected—and possibly might contain if not actual errors at least an account of circumstances lacking in some important detail. [If, for example, an inspector records that your premises has a wooden staircase—even if the staircase actually is cast concrete—from the point of view of the Inspectorate it will be 'wooden' always. (We write of a known occurrence.) Subsequent inspectors will ask to see your wooden staircase (presumably to check its condition) and will be too polite actually to say they do not believe you when you say there is not now and never was such a thing. Should you have the misfortune of a fire at your premises the wooden staircase mentioned in the file might well turn up in evidence against you.] It is desirable therefore to take time with the officials to go through the items noted and to ensure so far as possible that they are accurate.

One must have all the permissions required for every activity in which one is engaged and be punctilious in informing all the local offices concerned of every change in circumstance. Especial care should be taken over the drafting of the letters concerned (all of them will be kept on file), so that they give precisely the information you wish to convey and bearing in mind that they may be produced in evidence at some time in the future—when they will be considered in public by people who very likely will not be disposed in a friendly way towards you. The contents of such letters should be expressed in simple terms, and should be factual, informative, and realistic. The letters should be polite in tone and suggest no elements of anger, resentment, or disrespect.

Should an inspector consider you have not made a suitable and sufficient assessment of a hazard he (or she) will serve you with a Notice to do so within a fixed period of time. In more serious circumstances he could order the closure of your business. You should remember always that somewhere—not very far away—there is at least one weighty dossier ready to be brought down on you should anything go wrong.

Case Law

At the time of writing the newspapers are reporting frequent prosecutions under various aspects of health and safety legislation but it is too early as yet to be able to say with any confidence what interpretations might be placed in the courts on phrases such as 'substance hazardous to health', 'a micro-organism', 'suitable and sufficient', 'all reasonable steps', and 'suitable procedures', as used in the Regulations. There has been so far a noticeable willingness to plead 'guilty'—perhaps because of the expense of fighting such cases, and possibly because of the unfortunate publicity that might follow—and this has helped to leave such interpretations open, except in the senses in which they are defined in official 'codes of practice', 'guidance notes', and so forth.

The Regulations include a paragraph on 'Defence', which reads as follows:

> *Defence in proceedings for contravention of these Regulations*
> 16. In any proceedings for an offence consisting of a contravention of these Regulations it shall be a defence for any person to prove that he took all reasonable precautions and exercised all due diligence to avoid the commission of that offence.

In other words, the onus would appear to be on the accused to prove his innocence and not upon the authorities to prove guilt. From the point of view of the latter it could well be enough simply to show that an incident took place, but the accused will require time and resources to marshal the documents and to brief solicitors or counsel with a view to ensuring that his defence is presented in a satisfactory manner. However, besides the trouble and expense involved in all this he needs, before deciding what course to take, to consider carefully local attitudes—perhaps labour relations are indifferent or poor in general, there may be hostility towards him from local residents and newpapers, even perhaps a willingness on the part of local courts to uphold local officials whenever they can. No one likes to accept quietly an accusation of a criminal offence, but even should 'other things be equal' and he is satisfied that he is on 'a level playing field' a certain resolution will be required to respond effectively.

With thoughts such as these in mind, readers are advised not only to have ammunition of the right calibre ready at all times, but to be prepared if necessary to fire it to good effect.

No doubt other factors will influence willingness to defend—age, family considerations, whatever they may be—in addition to those indicated above. One of the most important will be the degree of commitment of the persons who are accused. In the simplest of terms, at first sight it would seem unlikely that a manager in a large official establishment (say, a polytechnic or a laboratory owned by the Department of Trade and Industry) would be so concerned by prosecution as would the proprietor of a small business—the managing official knowing that his Authority or Department can defend him (tax-payers ultimately, supporting both prosecution and defence), while the businessman is alone entirely.

Preparing Your Defence

It could be said with justification that the entire content and value of this book is that it suggests ways in which an employer or manager can begin now to prepare a defence in case it should be needed at any time in the future. Some of them (perhaps even the majority) may never need it; some may decide it hardly is worth the trouble to defend themselves; while some may need defence desperately. Unfortunately, at this stage it is impossible to foresee in which of these categories a particular person may find himself. Perhaps the overwhelming reason for starting on preparations now, and for maintaining them in future, is not only the citizen's natural wish and willingness to comply if possible with each and every law, but also the fact that no one ever can be certain that such a defence will not be needed.

A comparison with motoring offences was mentioned earlier. It can be assumed that as with such offences the resources made available to support prosecutions will be increased each year—and so also, in the same way and in a direct relationship, will the number of convictions.

We commented earlier on the management element and on problems associated with finding out even what substances are present within different locations and individual units of an organization. The attitude of management on a matter such as this is of key importance and suitable methods are required to bring together all sub-managers and employees so that not only do they appreciate its importance for compliance with the Regulations but they take an active part—maintaining records and advising the officers concerned of all relevant changes in purchasing policy or working practice at their units.

It is to be hoped that legal sanctions will encourage more mature conduct in future but we have noted in our industrial work certain attitudes of mind and practices which militate against effective effort in questions such as these. An example would be the designated safety official who took an unduly humble stance and was too shy (or excessively conscious of his own lowly status) to insist that all employees must comply (the Regulations apply equally to the Chairman and directors, as well as to the messengers and clerks in the Post Room). There is an element of good public relations in requiring all visitors to a building site, factory, or quarry, to wear hard hats or other prescribed headwear and protective clothing but it is now also a legal requirement which must be enforced. A manager who is too servile to insist on this without exceptions must be replaced by someone more determined.

It is a truism that the art of effective management rests on somehow inducing staff to work together, but it is true also that many organizations are riddled with politics. (The Production Department might despise Research, everyone might think Management Services a waste of time, old employees hold graduate trainees in contempt, and so forth.) In such situations memoranda can be ignored and it can become a point of honour to raise objections rather than to comply even with simple requests. Officers responsible for enforcement cannot ignore situations like these and must find ways round them. It will not be enough just to send out letters and other documents: they will have to go to see the staff involved at their own offices, to talk with them, and to establish mutual respect in order to get the necessary work done.

Employers are permitted either to display an official poster giving a summary of health and safety legislation, together with the address of the local office of the Executive, or to distribute to employees leaflets with similar information. Both poster and leaflets suggest to employees that if they should think that conditions or practices at the place of work might harm health they should inform first their supervisor or safety representative. There is however at least an implied invitation to approach the Executive, and each year a proportion of employees (and ex-employees) does so. In some of these instances it could be suspected perhaps that the underlying motive is not so much concern for health and safety as to express a grievance but nevertheless each complaint will be investigated and the employer concerned could well find, with no warning, that all his activities (not only the one complained about) are under scrutiny from the point of view of health and safety. Maintaining good relations with all members of staff (and not forgetting any members of the public who happen to live nearby) is well worthwhile.

There are other ways in which it will be helpful to make the effort to change attitudes of mind. People of all ages have a certain propensity to 'play the fool'—perhaps to enliven a dull Friday, at Christmas, or before someone's Wedding Day—and boredom or (say) lunchtime drinking can bring out conduct of this kind. Virtually anything to hand can become involved when employees 'lark about'—a compressed air line, pepper, acetic acid, washing-up liquid, more searching cleaners, or worse—and such objects or substances might be thrown around, introduced into clothing, added to tea, and so on. (A former worker in the asbestos industry described how young operatives would play with the fibres, making from them false moustaches and wigs to wear.) Most Standard Conditions of Employment would include rules directed against tomfoolery but now more than ever it is essentially to make staff aware of the dangers and to ensure that they stop it immediately.

Another problem for management, related to the foregoing in that it calls for a similar awareness and sense of responsibility, is that of deliberate sabotage. Acts of this nature can be performed, without apparent warning, by employees of any age, sex, or standing in the organization—from the lowest junior to a member of the board of directors. Prevention probably is not possible, but reasonable steps can be taken to keep such risks to a minimum—as examples, by limiting access only to those areas in which an employee is required to work, prohibiting access outside working hours, and holding materials which are stocked in any quantity in locked cupboards or enclosed storage units. It also is important, as with 'larking about', to warn employees of their common interest in preventing deliberate adulteration of products, contamination, or damage in any other form. Once again, sending a memorandum or posting a notice probably will not be sufficient and it will be desirable to talk with the staff concerned and to impress on their minds the possibility of tragic consequences.

Accidents

Having improved communications with his managers and employees, and after warning them of the dangers inherent in playing the fool, the enforcement man (or woman) might turn attention to risks of accidents. Either he personally or other suitable personnel should visit all units in the organization and study them in detail, with the assistance of the managers and staff working there. Adequate notes should be taken, recording the dates and times of the visits, activities examined, staff concerned, and so forth. Any records of accidents in the past should be reviewed, and the effectiveness of remedial action assessed. In a particular unit it may be fairly easy to imagine how accidents might take place, but the assessor must be prepared also to estimate their likelihood and to ensure that enough corrective steps are taken.

As an example, let it be assumed that a small office machine has been placed in a room in such a way that it is close to an inward-opening door; further, at intervals there is a need to service the machine by changing a container of solvent or of toner, and the fitting of these is done in a cramped space close to the door. Replacing a single container takes only a short time; it would seem both practicable and advisable to issue an instruction that it should never be done unless a second person is present to protect the operator by preventing anyone from trying to open the connecting door while the change is being made. A parallel precaution would be to ensure that all windows were open and other ventilation turned on while the substances concerned were being replaced.

A very frequent form of accident which can result in a wide range of injuries and involve many different substances is that of an operator falling on a slippery, damaged, or uneven floor. The unfortunate individual concerned might hurt himself, or scatter whatever he is carrying—perhaps injuring others, damaging or contaminating equipment and goods. With this type of accident in mind it is sensible to inspect floor surfaces at frequent intervals, and to give instructions that all spills (whatever substance might be involved, including water) should be notified immediately. Particular attention should be devoted to the suitability and condition of the materials from which floors are made and to features such as whether they are flat, the methods of joining and of fixing them to the substrate, occurrence of steps or other unevenness where levels or floor finishes alter, and so on. The cleaning preparations used should be 'non-slip' and every effort should be made to prevent the spilling of substances of any description (including solids as well as liquids, rain, snow or mud, coffee, tea, squashes, and fruit juices as well as oily liquids), besides making certain that if a spill should occur it is marked off and cleaned away both quickly and effectively. (It is not sufficient simply to display a notice warning employees to take care.) For similar reasons, when concrete or mastic floors are painted, they should not be finished to be smooth and shiny; the high polishing of wooden floors should be avoided too.

In familiarizing himself with the offices, factories and other facilities of his organization the assessor may well note numerous possibilities for accidents involving substances, some of them perhaps seeming rather likely to occur (that is, to which a high degree of risk might be attached) and some perhaps far-fetched (though conceivable). He might see, to mention one or two typical examples, that drivers of lift trucks are treating chemicals shipped in polyethylene flasks with much the same amount of care (or lack of it) as if they were in steel drums, or that substances packed in plastic film bags have been stored near to windows in strong sunlight. He may need to ensure that handling staff are aware of the rated capacities of cranes and hoists (nowadays these should be expressed in metric units, unless it should be that items to be lifted and carried are consigned still in quantities such as 'hundredweights' and 'long tons'), and even to calculate and to demonstrate to the drivers how many sacks, bags, drums, or flasks the lifting capacities represent.

Many facilities—both offices and factories—simply have grown up over a period of time on sites which even originally were not particularly large or convenient: the parcel of land may be awkward in shape (say, long and narrow), be made up from several small areas, divided by public road, railway embankment, river, or disused canal. The patterns of internal roads, services, and pipe-lines may be rather illogical and troublesome as a result. The assessor needs to help ensure that deliveries of supplies of all kinds can be made safely at all times into the correct stores, tanks, or silos, and that delivery and distribution vehicles move within the unit in a well-ordered and satisfactory way. (It may be necessary, even, to have some vehicles designed and built especially for use within the narrow roadways and limited headrooms on the site.) Numerous prescribed drills and warning systems may be required, so as to protect junctions and safe working of 'one-way' routes.

Long-standing practices at the point of delivery may need attention: as an instance, many units buy supplies of gases under pressure in heavy metal cylinders and it is not advisable (as has been seen occasionally) that the cylinders be rolled off the back of the delivery vehicle and allowed to fall under their own weight to the ground or pavement. The cylinders should be treated with reasonable care and suitable hoists always used.

The use of contractors for removals and to maintain or modify buildings and services requires careful planning in advance; any changes to safety systems which are necessary because of the work should be supervised personally to see they are functioning as they should.

Normal Working

It is necessary to assess risks of all the types envisaged (including the possibility of accidents) but the primary object is to ensure safety under conditions of normal working. Having secured details of all substances in use at the unit (including data on toxicology) he or another competent person should examine each operation in which the substances are used and in due course make the assessment required.

The current edition of 'The Merck Index' (a standard reference) contains details of 10 000 substances, no doubt all of which could be taken in some degree to be 'hazardous' in some circumstances (as also could be—say—tea, coffee, or water). Many of the substances are obscure and not likely to be encountered in most places of work, but all of them will be found somewhere, and sometimes also in a variety of combinations and different formulations.

In most situations—as at a small factory—the number of different substances to be assessed may be little more than perhaps a few hundred, and fortunately for the assessors (and, bearing in mind the costs of the work, for their employers) the really large numbers of substances are to be encountered in more specialized locations—such as laboratory suppliers, hospital stores, manufacturing chemists, pharmacies, research establishments, retail shops, universities, colleges, and schools.

Again, keeping to the simple case, a small factory may well have comparatively simple operations—such as merely packing a restricted range of a game such as draughts ('chequers', or 'checkers') so that assessments may be for a few items such as:

(i) paper dust
(ii) wood dust
(iii) pathogenic micro-organisms
(iv) glue residues
(v) decomposition products from plastic film
(vi) cleaning compounds and their decomposition products
(vii) contamination from sites nearby.

Pharmaceutical factories are controlled and licensed under separate arrangements but a typical larger unit manufacturing (as an example) common flooring materials easily could require assessments for more than a thousand substances, bearing in mind all types of raw materials, colours, and other additives and compositions in the range. Even non-manufacturing organizations—such as large chains of banks, bookmakers, or building societies will find on investigation (if they do not know it already) that they buy and use large numbers of different things.

Every substance must be noted and identified, with details such as composition, brand name, form in which it is supplied, type of packaging, toxicological data, and so forth. Having established this basic documentation, the assessor should enquire into the ways in which each of the substances is used under normal circumstances. This may be easy (the substance concerned may be in regular daily use) or difficult (the substance may be used rarely, or even never—as, since St. Patrick banished snakes, one hopes would be the case with snake-bite antidote in Ireland). The assessor may take it that if a substance is held in store but not used the 'risk' from it will be slight, but on occasion an assessment of risk will be more difficult to make than this. It may well be worth checking, as instances, how frequently stocks are replaced, whether conditions of storage are as recommended and satisfactory, and how time-expired stocks are dealt with or disposed.

Even with substances which are used frequently or every day as a matter of routine there may be some difficulty in establishing precisely what 'normal working' is, and so assessing risk. Most manufacturers offer ranges of products and these can encompass many different formulations: which are the formulations that are used in 'normal working'? When two or more lines are being operated, which products should be on the two or more lines under conditions of 'normal working'? If one of the machines is out of use for part of the time—is that 'normal'—and if one or more of the operators is away on holiday or ill? Other factors which should be considered include changes in ambient conditions (are the windows closed and the ventilation turned off in winter?), whether the unit is on the route to the canteen (with other workers passing through frequently, possibly carrying food or smoking), and whether decorators or maintenance men often are in there with ladders and other equipment.

We think it worthwhile that an assessor should listen not only to the manager of a unit but also to the operators, including shift workers if there are any. It may be just that some members of the staff like to grumble but also it may be true that fumes accumulate in the course of a week, and are worse on Friday than on a Monday morning. It also could be worth finding out what happens in practice during the night. (Do the operators really run the equipment a few hours at higher temperature and double speed, then close down for a break? If so, what are the implications? Can you hope to stop it or should you accept that it is bound to happen?)

Things Unexpected do Occur

An assessor may feel his work is complete and that he has made every allowance he could be expected to make for variations in 'normal working'. The office, shop, factory, or laboratory concerned is well-run, everything is documented, proper precautions have been taken, silly practices have been stopped, and he can devote more time to keeping the records and procedures up-to-date.

Unfortunately, though, his work is not complete. He needs also to try to ensure not only that he himself is ready to cope with an unexpected event as soon as it occurs, but that in each sphere of an organization's activities there will be a competent person alert to the possibility of hazard.

We can give disguised examples to help illustrate this. Let us say that a project requires the development of land which has been used previously for other purposes; the results of analyses of trial borings and of other samples (including surface water) taken from the site could well show that the construction men would be in an environment containing quite a number of substances for which assessments had not been made.

Local authorities are establishing registers of land said to be 'poisoned' but it is quite possible that for some time in the future the registers will not be complete; even if records of previous uses of a piece of land do exist they may not show all the various substances which have been leached down or buried there. (In a few instances, descriptions given might have been worded carefully so as to be imprecise.)

In a mature industrial society on a crowded island some rather unexpected contaminants of land can be found. Assessors should consider not only what information is available from the records but also the individual stages in the processes which are known to have been carried on—for example, if ores were worked, what impurities might have been present in them, and in what amounts? Considerable care seems to us always to be desirable in proposals for re-habilitating waste or former industrial land. It can be helpful to talk with older inhabitants of the district about activities in the past—particularly about standards of housekeeping at the site and the methods used to dispose of waste. Some public libraries keep collections of old trade directories and these too can be of help.

It may be that for some brief period (in, say, 1917 or 1939) an area of land had a military use (as examples, as a Territorial Army or Home Guard depot, or a transit camp)—with the possibility of ammunition or other harmful devices having been dumped or left there inadvertently. Former airfields have been used more recently as 'industrial parks' and occasionally government land and buildings pass from one Ministry to another (as from Defence to the Home Department for an internment centre or prison). Searching files of newspapers can be very time-consuming but occasionally a local historian or librarian will be able to assist by retrieving helpful details from the local press. Another possible source of historical or technical information in a particular locality could be older scientific staff at a nearby college or school. In due course it may be necessary to take specialist advice on questions such as the likely products of decomposition of the substances that were stored or used in the trade or unit concerned.

Remembering also that the Regulations include microbiological requirements, it might be just as well to ascertain whether a plot of land was (say) part of an old burial ground or the site of a mediaeval fever hospital.

Just as extraneous substances may be found in the earth they can be found with equal or similar unexpectedness in water supplies or in the air. Industrial users of water find it necessary to test constantly for a limited range of parameters (colour, alkalinity, *etc.*) to ensure the suitability of the water for the work in hand; occasionally more rigorous investigations are necessary to overcome problems.

Water is used for a greater variety of purposes than any other substance, and the specification or standard of quality it must meet can vary a great deal depending upon the use in question. Some of the highest standards are set for water to be used for pharmaceutical purposes, with water for drinking ('potable' water) not far behind. The quality of water is a common source of complaint and of difficulty at drinking fountains and in vending machines. On some sites, drinking water may be discoloured, have odour, or even visible contamination such as black specks or flakes of rust. Even in more fortunate locations with clear water there can be complaints of elusive 'taint' or an unpleasant taste which may be very difficult to identify.

At times, rather surprising results can be found when swabs are taken for examination from drink vending machines—and we suggest it is essential to have such machines and drinking fountains checked at regular intervals for cleanliness and hygiene. (In this connection, it can be of value to listen with some sympathy to complaints about dirt or 'a funny taste' in the supply of drinking water.)

The quality of potable water also can be a factor in mass catering, for canteen meals, and so forth. Defects such as the presence of rust or taint can be transferred and become evident also in prepared foodstuffs such as soup, vegetables, and puddings.

There are further comments elsewhere on the monitoring of atmosphere at the place of work for dust or for more specific contamination but as a final illustration of the need to expect the unexpected we would mention an incident at a small unit in England which was re-packing for sale in retail shops items which were made in the Far East and received at regular intervals in bulk consignments.

The unit concerned was new, airy, and well-ventilated, and its usual operations presented little or no risk. (It is probable that on the whole site there were no more harmful products than, say, paint and washing-up liquids.)

Unfortunately, however, the exporter in Asia sent one of the consignments of goods in large board boxes which had been re-made. The outsides of these boxes were of plain kraft board but the interiors were printed in two languages with health and safety warnings which suggested that the material was second-hand and had been used previously to pack a powerful herbicide. A few members of staff at the re-packing unit started to experience symptoms, such as irritation of the skin, which were consistent with contact with a substance of this nature. All work was stopped and after an urgent investigation quantities of the herbicide mentioned in the printed warnings were identified not only on the re-made boxes but also on the manufactured goods sent in them.

In other words, because an exporter on the other side of the world tried to save money on his packaging a fairly rare and powerful irritant turned up without warning in an environment that otherwise was thought almost risk-free.

Assessors should try to ensure that there will be present always a responsible person who can deal promptly with a situation such as this. It is easy to say, but the action required in this case was to stop production immediately, ensure that operators washed thoroughly, discarded contaminated clothing, and—if experiencing any signs of irritation of the skin or of other illness—were examined by doctors. The person concerned had to have authority enough to close the factory and to send staff home—decisions both hard to make and to put into effect, not to mention the question of explaining them later should they be shown not to be justified.

Who Should Assess?

Some of the qualities likely to be useful to an assessor have been indicated already—ability to master documentation, patience, tenacity, powers of persuasion, *etc.*—but a point may be reached when it simply is unrealistic to expect more from the person or persons concerned and further advice and assistance are needed.

There is a role for the analyst in monitoring for substances, assessing risks, and developing analytical methods—say for substances in human or animal fluids, in tissues, collected from atmosphere on sampling devices, and so forth—and in providing objective data and advice.

The types of enquiry received are varied in character and generally involve the taking and analysis not so much of long runs of routine samples but rather discrete groups of items requiring a degree of individual treatment and even some novel considerations at times (such as devising appropriate methods of collection). A visit to a workshop or to other premises to assist in matters of hygiene usually invokes at least some elements of the forensic investigation: besides the immediate scientific aspects, the consultant should note the design and materials of the building, the arrangement of activities and the speeds at which they are carried on, the form of heating and ventilation, the number of operatives engaged, whether protective equipment and clothing are necessary (and whether they are being worn), the temperatures at which the work is done, the possibilities of contamination, what safety precautions are taken, and so forth. The eventual report will include not only results of the analysis of the samples taken, together with references to toxicology, but also appropriate comments with regard to the conditions seen.

We think it good practice at an early stage to visit the establishment and to talk with the operators before setting up equipment; they should be able to comment on the conditions under which samples are taken, on the sites selected for sampling apparatus, and it is preferable that they should be in agreement with what is being done. (The analyst may need to win the confidence of the employees and to overcome the suggestion in their minds that he or she is 'on the side of' the employer.) Full notes should be kept of the circumstances of sampling (including dates, times, and names of operators concerned), with sketch plans or diagrams if appropriate, and often it is desirable to make ancillary readings, such as of ambient temperatures, relative humidity, or (particularly when working out of doors) of the North–South axis, wind speeds and directions, mist and fog, emissions visible on sites nearby, and so on. Besides a compass, anemometer, and devices to measure distance, a good camera can be most helpful when it is possible to use such a thing (always after asking permission) without danger of compromising commercial secrets.

Financial considerations will be a factor in deciding the extent and complexity of any sampling programme; the client will wish to have value for money but it is in his interest too that the samples taken should be representative.

When sampling and analysis have been done and the 'customary' or 'typical' conditions within a unit so expressed in numerical terms there is a basis not only for assessment of these conditions but also of exceptional situations—as in leakage from a storage vessel or the possible effects of spilt material.

Various ranges of equipment are available for sampling and there is also a wide choice of portable meters and other instruments for 'direct reading'. However, our view still for work of this general nature is that it is preferable to return the samples to the laboratory for the analytical procedures and calculations. With proper care, some equipment may be transported but the setting up and calibration at the site, often for short periods of time only, can present difficulty. In addition to our preference from the scientific point of view we think that when this is possible there are sound psychological reasons for returning to the laboratory and preparing a written report with due care (rather than giving on-the-spot readings verbally or as 'print-outs', perhaps not altogether systematically, to excited managers or personnel).

No doubt other specialists—such as medical practitioners and members of professions such as occupational hygienists—will be engaged also with meeting requirements of the Regulations, and in this context the practical value of the Royal Society of Chemistry's 'Indicative Register of Health and Safety Specialists' will be of considerable interest. For our part, while we would not, of course, expect to be engaged in matters of medical practice neither would we wish it to be suggested that medical doctors might write analytical methods. On the other hand, although taking the samples is an important aspect of making the assessments it is likely that unqualified staff will be engaged sometimes for work of this nature, and for 'routine monitoring'. Some dangers could be anticipated and a chartered analytical chemist should be a person suitable for drawing up written procedures for the sampling and monitoring; not only initial training but also regular supervision, authorization, and approval would be desirable for scientific procedures of this nature, which are to be carried out in practice (perhaps in cramped or otherwise difficult working conditions, at remote hours of the day or night) by rather more limited and sometimes inexperienced staff.

A 'qualified person' is likely to be required if not for the every-day details of monitoring and sampling at least for establishing procedures and for checking at intervals that they are being carried out as they should. Staff analysing the samples received, whatever their character, preferably should be qualified and under the regular supervision of a qualified and experienced analytical chemist. Persons developing or modifying methods of analysis to deal with new problems arising under the Regulations certainly ought to be well-qualified and experienced in analysis. Finally, in the field of toxicology, in order to give advice on the safety of materials and in the expression of opinion about safe procedures, one would have thought that seniority and substantial experience would be mandatory.

Analysis of Environmental Samples

Samples may be taken or received in a variety of forms and will be subjected in the laboratory to an appropriate range of preparatory and analytical techniques—including examination by chemical and physical methods, light microscopy, infra-red spectroscopy, atomic absorption spectrophotometry, and chromatography of various kinds.

Some details of the use of light microscopy for samples of this nature are provided in 'An Introduction to Applications of Light Microscopy in Analysis', particularly in Chapters 12 and 13. Stated very briefly, various kinds of light microscopy can be used for the identification of components and of sources of contamination in many different types of sample.

Among other techniques, infra-red spectroscopy is used in identifying:

Asbestos
Cleaning agents
Contaminants in foodstuffs and arising from the packaging of foodstuffs and pharmaceuticals
Dyes and pigments
Plastics and polymers

Examples of the use of atomic absorption spectrophotometry (including flame emission, flameless atomization, and hydride generation where applicable) include determinations of the following:

Aluminium	Mercury
Antimony	Nickel
Arsenic	Palladium
Barium	Platinum
Beryllium	Selenium
Cadmium	Silicon
Chromium	Silver
Copper	Tellurium
Gold	Thallium
Lead	Tin

A selected list of substances and situations experienced and involving analysis by chromatography of various kinds is given below:

So-called 'acetone' (reclaimed solvent) used for cleaning
Acrylamide
Adhesives (particularly synthetic resin types)
Alcohols in atmosphere
Some anions
Benzaldehyde
Benzene
Butyraldehyde as contamination
Butyric acid
Carbon dioxide, disulphide, and monoxide (one instance of the latter involving deaths), and other gases
Cleaning compounds

Contaminants in foodstuffs and water
Copying fluids
Creosote
Decomposition products (of papers, pharmaceuticals, resins, tobacco, *etc.*)
Dioxin in incinerator emissions
Ethylbenzene
Ethylene oxide
Formaldehyde
Glycol ethers and derivatives
Hexanes (dimethyl and trimethyl), heptanes, octanes, and nonanes
Inks
Isocyanates
Isophorone
Monomers such as acrylonitrile, styrene and vinyl chloride
Oils (of various types, including cutting oils)
Ozone (in photocopying, welding, *etc.*)
Paints, and other surface coatings
Pentachlorophenol
Perchloroethylene
Permethrin (*cis*- and *trans*-), and other pesticides
Phthalates and other plasticizers
Photo-sensitive materials
Solvents and solvent vapours
Toluene
Trichloroethane
White spirit

For Further Consideration

The Regulations have been drawn widely and while the degree of effort necessary to ensure compliance will differ depending on the type of establishment it is reasonable to expect that in many cases substantial resources will be needed. The central problem faced by the assessor is that of reconciling given facts of the toxicology of a substance, or of many substances, with actual conditions of use at the establishment. Such a reconciliation is not and cannot be wholly a matter of fact: it is necessary also to make judgments about matters such as:

(i) reliability of procedures
(ii) degree of standardization of supplies
(iii) consistency of working
(iv) possibilities of intervention (such as failures in services)

—and somehow to take matters of this nature (including 'human factors') into account. All the relevant facts should be gathered, and the judgment related closely to these facts, but the element of decision cannot be eliminated; any decision can be proved wrong later by events.

Fortunately, some help is to hand—in the publications of the Health and

Safety Executive, from growing experience of the working of the Regulations, from consultants, from professional institutes, from trade associations, suppliers of business systems and information, and (not least we hope) from the present volume.

Another important source of assistance (bearing in mind some of the comments made earlier in this chapter) may well be your local office of the Health and Safety Executive. It is always desirable to be on good terms with staff engaged there and perhaps is worth emphasizing that they should not be seen as 'the enemy' but are ready to help with advice.

CHAPTER 2

A Viewpoint from the CBI

JOHN BANHAM

The Confederation of British Industry has long supported the production of regulations to control hazardous substances at the place of work, to protect not just the health of workers but also the health of all persons likely to be affected by work activities. The Control of Substances Hazardous to Health Regulations—the so-called 'COSHH' Regulations—are the outcome, substantial controls which put the 'health' into health and safety at work. They are thought by many to be just as important as the original Health and Safety at Work Act of 1974. They apply to every employer. They are onerous and will cost a great deal in management organization and time but the benefits should be at least equally great. This chapter discusses in turn the problem that the Regulations set out to tackle, the solution that has been agreed, and the next steps.

The Problem

As years have passed there has been a dramatic switch in the causes of death to the population as a whole. Infectious diseases, childbirth, infant mortality—diseases associated with poor hygiene—feature no longer in the statistics as serious causes of death. They have been replaced by heart attacks and strokes (50% of all deaths), cancers (25%), and respiratory diseases (10% of all deaths). External injuries from accidents and from poisoning (including incidents at work) account together for only 3% of deaths. The primary risks, therefore, are not associated with incidents at work but rather with factors such as lack of regular exercise, smoking, alcohol abuse, drug abuse, and inappropriate diet. Britain has the highest rate of deaths from heart disease in the world—some 180 000 deaths a year according to the

Royal College of Physicians. In contrast, in 1985 the Royal Society for the Prevention of Accidents reported all accidental deaths as follows:

General circumstances	*Accidental deaths*
	number
In home or leisure	5700
On the roads	5500
At work	500*

—and from this comparison of figures it would appear therefore most desirable to devote resources to fostering safety in the home and to improvements on the roads (this is one reason why the CBI has long campaigned for increased investment in the transport infrastructure).

But with hundreds of employees killed each year at work and up to 300 000 injuries sustained one hardly can be complacent about the record for occupational safety. Aside from immediate tragic consequences there is also a cost of accidents and of occupational ill-health, estimated in total to be of the order of £2000 million annually. In any case, society expects much higher standards of health and safety for people at work than for householders or drivers of private cars. Employers are expected to protect their work people more effectively than, admittedly under rather different circumstances, they appear to protect themselves. This premise applies as much to any risks posed by hazardous substances as it does to questions of the safety of machinery.

A discussion paper from the Health and Safety Executive on the difficulties of collecting information about illness or disease caused by work opens with the assertion: 'each year more people in the UK die from diseases caused by work than are killed in industrial accidents'. A statement of this kind encompasses illness caused by asbestos and by other substances where harmful effects may appear years after exposure, as well as from harmful substances and other factors in combination, from personal sensitivity, and from causes virtually indistinguishable from non-occupational ones. None the less, today's statistics reflect the conditions of previous years.

Table II overleaf gives the total numbers of selected 'dangerous occurrences' which were reported in each of the years 1981 to 1985. Most of these occurrences would have effects which were immediate or not long delayed: they should be considered in conjunction with intermediate and long-term effects of exposure to substances at work and deserve thoughtful attention on the part of managers.

A legal framework and institutional arrangements for dealing with hazardous substances have been in place since 1974, when the Health and Safety at Work Act established a comprehensive framework of controls and placed clear duties on employers to protect their employees. Some emphasis was placed on self-regulation and on controls that were 'reasonably practicable'. The Health and Safety Commission was set up to oversee application of

*Figures published by the Health and Safety Commission for 1988–89 showed 514 fatalities at work, including 167 from one incident—an explosion on a production platform ('Piper Alpha') in the North Sea; the total of deaths in the previous year was 360.

TABLE II *Selected 'dangerous occurrences' reported, 1981 to 1985*

Classification	*Year*				
	1981	1982	1983	1984	1985
Uncontrolled release or escape of substance or substances which could be harmful	843	671	626	576	684
Personal exposure to or contact with a harmful substance, or deprivation of oxygen	206	147	142	143	158
Others	2185	1881	1638	1687	1678
Total	3234	2699	2406	2406	2520

the policy, with its operating arm—the Health and Safety Executive—to prepare regulations under the Act, guidance in the form of publications, and to ensure satisfactory enforcement in practice. The Commission is a statutory body, with the authority to recommend legislative proposals in the field of health and safety at work to the Secretary of State for Employment and to other Ministers. The Secretary of State appoints the members of the Commission, after consultation with the Confederation of British Industry, the Trades Union Congress, and others. Nominations for membership of the Commission are put forward by the CBI, the TUC, and by local authorities; it is assumed that in its consideration, say, of new regulations, certain members of the Commission will represent and express the views of employers.

We believe that because of the 'tri-partite' nature of the Health and Safety Commission all aspects of proposed new regulations are debated thoroughly in advance and that they emerge from these discussions both workable and fair. Because of such debates new regulations do not invoke surprise when they are applied in practice, nor do they impose excessive bureaucracy on managers or employees.

However, in the course of the 'seventies it became increasingly clear that the pre-occupation under the Act was safety, and that a strategy for control of substances hazardous to health was lacking. Since 1980, in discussions with the Health and Safety Executive, members of the CBI have developed a philosophy and a six-point set of proposals in order to deal with hazardous substances at work, as follows:

(i) It was recognized first that the old legislation was confused; there existed about fifty different sets of regulations and orders in the field, covering hazards with products as diverse as biscuits and Siberian horse hair. It was accepted that there was need to continue, for example, certain parts of the Factories Act, duties existing under the Health and Safety at Work Act, and requirements of Common Law, but it was felt that modernization was an urgent requirement, with more specific control over substances of a hazardous nature.

(ii) In this field, as for business in general, there was need for a simple and modern legal framework which would make the responsibilities clear.

Revision would permit the repeal of existing legislation which met present-day needs no longer, and could reflect, codify, and formalize up-to-date good practice in progressive organizations.

(iii) From the point of view of good relations with the public at large it was considered desirable to show that employers supported high standards in the control of substances at work, and also the enforcement of high standards through new regulations.

(iv) Good practice in health and safety was desirable and legislation would ensure uniform application and enforcement.

(v) The chemical industry in particular held the view that there was a need for a single set of regulations covering all work with substances hazardous to health (not, as previously was the case, just some substances in some situations).

(vi) The European Council of Ministers proposed revision of the framework Directive on protection of employees against risks related to 'exposure to chemical, physical, and biological agents at work', and more directives relating to substances were being brought forward; it was desirable therefore to clarify the position in Britain and be in a position to influence more effectively the form of new controls—*e.g.* overexposure to carcinogens.

For these reasons, the principle of a comprehensive set of modern controls was accepted, and it remained only to try to ensure that they could be understood, were workable, and fair.

The Solution

The Control of Substances Hazardous to Health Regulations 1988 were introduced under the Health and Safety at Work Act of 1974 and it was within the forum of the Health and Safety Commission that many of the negotiations—certainly the most fruitful of them—were carried out. Much credit is due to the current chairman of the Commission, Dr John Cullen, for holding the ring so well and ensuring that employers and employees never lost sight of their common objective of providing better protection for all affected by hazardous substances at work.

The Regulations apply to all substances and micro-organisms which are hazardous to health, and to dust of any kind when present at a substantial concentration in the air. The employer is required to:

(i) Assess and review systematically the risks to which employees and others might be exposed through contact with, or inhalation of, substances and dust at the place of work.

(ii) Ensure that the exposure of employees to a published list of substances is controlled below prescribed limits, and that exposure to all substances is controlled to the extent needed to prevent adverse effects on health. Measures for control should preferably be through systems of work, design, and engineering, and as a last resort through personal protective equipment.

(iii) Maintain engineering controls (such as the extraction of gases or solvent vapours) in an efficient manner and monitor both conditions at the place of work and the health of employees to the extent necessary to confirm that exposure is being controlled adequately.

There are commensurate duties on employees to use protective equipment if it is provided, and to report defects, but overall responsibility rests with the employer to make arrangements to reduce any risk of damage to health, and he will have to suffer legal consequences if such arrangements are judged to be insufficient. (Publication details for the Regulations and for codes of practice, guidance notes, and other relevant material from the Health and Safety Executive, which are essential reading, are summarized at the conclusion of this chapter and are presented more fully in the references on page 33.)

Diligent efforts were made to obtain regulations which would be acceptable and workable, and these included regular, detailed discussions with members of the Health and Safety Executive and the trade unions: gradually, there emerged twelve criteria against which the value of the final text could be evaluated, as follows:

1. The Regulations should be simple enough and expressed so as to be understood readily by members of organizations of all kinds, large and small.
2. They had to be practicable, so that they could be applied by farmers, by manufacturers of chemicals in bulk, in heavy engineering, on building sites, in offices, shops, laboratories, waste disposal, and all the other diverse enterprises—from vast corporations to two-man garages.
3. They had to be cost-effective.
4. They had to reflect good management and professional practice, not penalizing companies that had achieved this already.
5. The degree of stringency of control had to be related to the seriousness (or otherwise) of the risk—to provide a proper rationale for action and, indeed, for enforcement.
6. They had to be flexible enough to accommodate changes which might be needed in standards of control in the future, should unsuspected or underestimated hazards come to light, and to avoid imposing obstacles to the use of new technology and control techniques.
7. Sufficient flexibility was needed also to enable users to meet without need for further legislation on the requirements of further directives on the protection of health which are being negotiated currently or will be negotiated in the future.
8. Similar flexibility had to be inserted also into the Health and Safety Executive 'Guidance Notes'—as examples, those on occupational exposure limits (EH 40), Monitoring strategies for toxic substances (EH 42), and Substances for use at work: the provision of information, 1989 (HS(G) 27).
9. Discussion with these objectives in view had to include representatives

of the Trades Union Congress and professionals in the field of occupational hygiene.

10. General guidance on assessment had to be agreed and published in simple form well before the Regulations were brought into force.
11. There was a need for special guidance to be written by joint teams to help the building industry and those engaged in agriculture to understand and to use the Regulations.
12. A prompt start was needed to work in the Commission's Advisory Committee on Toxic Substances to review twelve substances in Schedule I of the Regulations which were felt by employers not to present a high degree of risk, together with certain substances in the Carcinogens Code not thought to be appropriate to it.

It is difficult to judge whether these criteria were met fully. Certainly the Regulations relate controls to degree of risk and in that sense are cost-effective. Nevertheless, even for alert companies which operated to high standards, the costs of re-structuring management systems—and especially of installing systems of record-keeping intended to prove conclusively compliance with the Regulations—have been high, whilst the health benefits may not have been improved greatly. Certainly there is flexibility in the system, but this in turn has placed on trade associations a considerable burden in outlining common standards for their sectors. It has led to a plethora of guidance material which may have served to confuse. Also, there have been misleading statements about the actual requirements of the Regulations, sometimes with a commercial motive.

So whilst employers called for flexibility in practice this brought extra demands which are not always easy to accommodate or to understand. Throughout the debate on the Regulations employers stressed the relevance of a sound cost/benefit analysis, since heavy costs would be incurred. For example, the activities of British Steel, apart from steel-making, range from the production of gas and coke to purification of water, and it uses in these more than 10 000 substances: it estimated that including the multiple use of substances in integrated processes it would need to prepare more than 30 000 COSHH assessments.

The high costs involved would be acceptable if there were a commensurate benefit to health in the long term. Ministers and Parliament should realize that the total package will represent a considerable international cost handicap, particularly for engineering and general manufacturing, construction, farming, quarrying, and many other activities (industrial and otherwise) that will have to assess the use of thousands of substances in a wide variety of operations. The costs of preparing for and implementing COSHH now are becoming apparent, as are the savings to be made by designing projects from their inception with health and safety considerations in mind.

The critical decisions were taken by the Health and Safety Commission at meetings on 8th March and 22nd March 1988 when final drafts were approved for the Regulations and Approved Codes of Practice. The Commission recommended that the Secretary of State for Employment bring the

Regulations into effect from 1st October 1989, apart from regulations 17(1) and 17(4), which would be brought in later, on 1st January 1990. The Commission agreed further that all sectors would have the opportunity to write notes of guidance or codes of practice, whether through industry advisory committees or other arrangements. The Executive would start to review urgently, through established machinery, specific substances identified by the Confederation of British Industry as being inappropriate for inclusion in the Regulations.

The Chairman of the Health and Safety Commission sent the texts to the Department of Employment on 11th May 1988. In July of that year the Secretary of State sought the agreement of other interested departments—such as Trade and Industry, and Agriculture, to his placing the Regulations 'on the table' in both Houses of Parliament. They were brought into effect in October 1989 as recommended by the Health and Safety Commission, apart from the provisions for assessment—which were held back until January 1990 to allow time in which management systems could be put in place.

Next Steps

Early in 1990 employers began the most difficult stage in the bringing in of the new Regulations: the Commission passed to the Executive, to employer organizations, trade associations, and industry advisory committees the responsibility for working out how to implement COSHH in every place of work, sector by sector.

Providing guidance on this has involved a great deal of work on the part of the organizations concerned. Enterprise, imagination, and foresight were required in planning the sector guidance, besides energy and application in writing it. However, the papers concerned should go a long way towards solving the problems. While insisting on sensible, practicable, and reasonable procedures, reflecting existing good practice, employers are determined to maintain high standards of protection. They wish also to be seen to be achieving high standards of care for employees, while continuing to provide opportunities for employment in the face of strong international competition (sometimes from countries in which there are different attitudes towards health and safety in general as well as towards the enforcement of regulations).

The Regulations for Control of Substances Hazardous to Health come as close to giving full protection against harmful substances at work as efficient managements are likely to be able to achieve. Looked at in this way, they have been successful in providing a framework. Even now, that framework is being built upon. Amendments to the Regulations, accommodating scientific advances and European legislative developments, are under discussion. Members of the business community play a full part in these talks, providing expert advice to many working groups and to the Health and Safety Commission on decisions of policy.

It now remains to be seen, over a period of many years, whether practice and the enforcement of the Regulations will match the care that went into their preparation. An important contribution towards achieving good results in this respect will be that of regular information and publicity material—which will help to inform employers of their obligations and of the positive opportunities that go with them. Without this, the opportunities that the Regulations offer for all concerned will not be realized fully.

References

'Health and Safety Statistics, 1985–1986', London, HMSO.
'The COSHH Package' (all published by HMSO):
SI No. 1657. The Control of Substances Hazardous to Health Regulations 1988.
Introducing COSHH (a brief guide for all employers).
Introducing Assessment (a simplified guide for employers).
Hazard and Risk Explained.
COSHH: Assessment (a step by step guide).
COSHH an Open Learning Course.
The Control of Substances Hazardous to Health Regulations 1988, general Approved Code of Practice on Control of Substances Hazardous to Health and Approved Code of Practice for the Control of Carcinogenic Substances.
Approved Code of Practice on Control of Substances Hazardous to Health in Fumigation Operations.
Approved Code of Practice on Control of Vinyl Chloride at Work.
Approved Code of Practice for Potteries.
Occupational Exposure Limits 1990. (Guidance Note EH 40/90, and subsequent editions).
Monitoring Strategies for Toxic Substances. (Guidance Note EH 42, revised 1988).
Substances for use at work: the provision of information (HS(G) 27).
COSHH Guidance for the Construction Industry.
COSHH Guidance for Universities, Polytechnics and Colleges of Further Education.

CHAPTER 3

Legal Requirements

S. G. LUXON

Introduction

Prior to the Second World War the greatest emphasis on protection of employees at the place of work was in the field of industrial safety. Since that time, with a rapidly expanding knowledge of the possible adverse effects of certain chemical substances and the consequent horrifying portrayal by the media of the diseases they can produce, there has been a universal demand for better control of such hazards and for more stringent legal requirements.

In the last ten years this has been overlaid by an appreciation of the long-term adverse environmental effects that certain chemicals can produce and even more recently by the publicity given to their disposal. Demands therefore have arisen for an integrated approach to such wide-ranging problems but thus far little progress has been made in putting this 'cradle to the grave' concept into practice.

Such considerations have led also to a wider appreciation of the problems associated with the production and use of chemicals, without which civilization as we know it could not exist. The social conscience of the population worldwide has been aroused in particular by the difficulties in striking a balance between protection and economic well-being for example, where the existence of certain communities depended on exploitation of harmful substances.

The subject has become one of national and international importance. The overall aim has been to ban, where alternatives are available, the use of certain extremely hazardous chemicals either because of their irreversible effects on the workers or because of catastrophic effects in the environment—carcinogens and chloro-fluorocarbons are examples. In the case of other hazardous chemicals the aim has been so to control their escape into the environment as to ensure that no significant risk to health arises.

As noted above, the principal result thus far has been directed towards greater protection of the workers and one consequence was that in Britain in 1974 the Health and Safety at Work Act was enacted. Its main purpose was to provide one comprehensive and integrated system of law dealing with the health, safety, and welfare of workpeople, and the health and safety of the public as affected by work activities. The Act has been described as the most significant statutory advance in the field of health and safety at work since Shaftesbury's Factory Act of 1833. It aimed to change radically not only the scope of existing law but also the way in which provisions for health and safety were enforced and administered. Thus it extended obligations and protection to five million or more people who never before had come within the scope of this kind of legislation—including workers in health and education establishments and research laboratories. It covered also for the first time the self-employed. The Act did not seek to cover every eventuality nor does it try to spell out rules for each and every work situation. It is an enabling instrument, the foundation of which is the concept of a general duty of care for people associated with work activities. Its very flexibility in this respect will mean that on the statute book there is a piece of legislation capable of being changed, expanded, and adapted to cope with risks and problems in industry for generations to come.

The general umbrella of the Act provides for an interaction of responsibility for the individuals and organizations associated with work or touched by its immediate consequences. The employer has a duty in law to his employees with regard to their safety and health, and those employees in turn have a duty one to another; so, too, the self-employed person has a duty to other people around him.

There are two significant advances which go much further even than the spirit of previous legislation. The general public now is entitled to a duty of care in terms of safety and health by people carrying out work activities, so that an employer, for example, not only has to ensure that his workers are safe, but also that members of the public who might be affected by any hazards from work activities are protected. The legislation includes an innovation: namely that an employer carrying on an inherently dangerous business or one that might be a threat to health if something went wrong, can be obliged to inform not only his employees but also the local population. There also are two specific requirements in the Act which are relevant to the subject matter under discussion. Firstly, it is the duty of anyone who manufactures, imports, or supplies any substance for use at work to ensure so far as is reasonably practicable that the substance concerned is safe and without a risk to health when used properly. The importer or supplier must carry out such tests and examinations as may be necessary to ensure this. He must provide adequate information about these matters and about any conditions or precaution necessary to ensure that the substance will be safe and without risk to health under correct circumstances of use.

Secondly, there is a requirement placed upon any person who manufactures a substance for use at work to carry out any research necessary with a view to the discovery and either the elimination or minimization of any risks to health or safety to which the substance may give rise.

In summary, the effect of this important Act was to extend the scope and coverage of earlier Acts both in respect of those it was sought to protect and in the range of substances to which it applied. Most importantly it provided for the making of Statutory Instruments which would put flesh on the bones of the Act itself. The 'COSHH' Regulations are of course one such instrument.

The Act not only broke fresh ground in its scope but provided also a new mechanism for giving practical guidance in respect of the observance of the legislative requirements—that is 'Approved Codes of Practice' which have special legal force. Such codes will not themselves be statutory requirements, but may be used in criminal proceedings as evidence that a statutory requirement has been contravened. If a person cannot show either that he observed the provisions of an 'Approved Code' or that he did something which would be as effective or better than the code recommended, he will be held to have contravened the statutory requirement to which the Approved Code was relevant (that is to say, one of the general obligations). (There is a similar relationship between the Highway Code and the Road Traffic Act.) Such an Approved Code of Practice has been made in respect of the Regulations for the 'Control of Substances Hazardous to Health'.

At the same time as these national initiatives were taking place the Commission of the European Communities was considering at a European level harmonization of the requirements in member states for protection of persons at work. This review was based on two sections of the Treaty of Rome. The first is a requirement to improve social conditions in member states and the second is the removal of barriers to trade. In this latter respect the indirect economic addition to costs associated with the control of hazardous substances must be considered part of manufacturing costs and hence will be reflected in the price of the final product. Differing standards in member states therefore would give rise to differences in costs and hence provide an indirect barrier to trade. In addition, with different legislative standards products might be refused entry into another member state because they did not comply with the relevant national legislation. In the light of all this the Commission published in 1980 a Directive designed to protect workers against physical, chemical, and biological hazards, and while many of the requirements contained in this Directive could be said to be met by the general provisions of the 1974 Act the opportunity was taken to tidy up the large number of individual regulatory requirements applying to hazardous substances, some dating back to 1906, by making a new comprehensive set of regulations applying to all hazardous substances coupled with an Approved Code of Practice giving guidance on implementation of the legislation.

Background

The 'COSHH' Regulations set down in a formal legislative frame accepted good industrial hygiene practices which have been observed by enlightened firms for many years. The principal additional requirement is to set down formally in writing procedures which have been practised over the last decade—such as assessment of the risks to health, measures to control the risk, maintenance of controls, monitoring where necessary, and health surveillance, instruction, and training.

Over the years, as the identification and understanding of the risks to health were documented, the discipline of occupational hygiene developed. Since the turn of the century health risks were identified in certain industries where hazards were appreciable, such as those using lead and its compounds. As a result, specific regulations were enacted to deal with the risks involved, and this led as time went on to a plethora of legal requirements with a concomitant need for regular updating (a time-consuming task).

Similarly, as indicated in the Introduction, action to control such hazards was taking place within the European Commission—resulting in the adoption of the so-called 'Framework Directive' on the protection of workers against chemical, physical, and biological hazards.

The Regulations

With a view to meeting these requirements, the 'COSHH' Regulations, together with its accompanying Approved Code of Practice, attempted to encompass the whole field of hazards to health from all chemical substances. The Regulations apply to every place at which persons are at work as defined in the Health and Safety at Work Act 1974 and where any person is exposed to a substance hazardous to health.

Such substances may arise in a variety of ways. They may be used in industrial processes; they may arise naturally, as in moulding hay; they may be used in cleaning or sterilizing; they may arise from the use of decorative materials such as paints, or may be produced as by-products.

The Regulations lay down procedures for controlling exposure to such substances and so protect those who otherwise might be exposed to them. They apply to all substances hazardous to health except those subject to recent legislation, principally asbestos and lead. Failure to comply with the Regulations constitutes an offence and is subject to the penalties set out in the Health and Safety at Work Act 1974.

Approved Codes of Practice

As mentioned in the Introduction, practical guidance on the Regulations is given in an Approved Code of Practice made under S16(1) of the 1974 Act. Such codes of practice have a special legal significance. It must be emphasized that although failure to comply with the Code is not an offence in itself, unless the person who is charged with contravening the regulation to which the provision in the Code relates can show that he has complied with the regulation in some other way, he may be deemed to be in contravention of the legal requirement. Thus, save in exceptional circumstances, the Code of Practice and Regulations should be considered together.

Prohibitions

Mention has been made of the need to tidy existing legislation in this same area and thus there are specific prohibitions on the import and use of certain carcinogenic substances (*i.e.*, 2-naphthylamine, benzidine, 4-aminodiphenyl, 4-nitrodiphenyl, their salts and any substance containing any of these compounds in a total concentration exceeding 0.1%, together with, for historical reasons, a ban on matches made with white phosphorus).

Substances Hazardous to Health

One of the keys to the application of the Regulations is the determination of what substances are hazardous to health. Unfortunately, no specific definition appears in the Regulations. Some guidance is provided in the 'Classification Packaging and Labelling Regulations' where parameters are set out for determining and labelling substances which are toxic, harmful, irritant, or corrosive. The CPL Regulations list many substances which fall into these categories. For substances not so listed, the parameters set out therein give a useful guide as to whether or not they might be included in one of these categories of hazard. Other useful (and more specific) information is set out in the Health and Safety Executive 'Guidance Note EH 40' which lists occupational exposure limits for some commonly-used substances. Substances appearing in any of these lists, and those having similar properties, must be considered as coming within the scope of the Regulations.

Assessments

The formal written assessment of the health risks created by work involving exposure to substances hazardous to health is the crux of the Regulations. Except in cases where the risk and the necessary control measures are obvious, this assessment must be in writing. The legal requirements may be summarized as follows:

Regulation 6 requires that an assessment be made of the work activity—

(*a*) to evaluate the risks to health arising from work hazardous to health, and

(*b*) establishing what must be done to meet the requirements of the Regulations, and

(*c*) the assessment be 'suitable and sufficient' in relation to (*a*) and (*b*)

(*d*) the assessment must be reviewed and, if necessary, brought up-to-date if any of the circumstances of the work change or if it becomes apparent that the original assessment is valid no longer.

The duty of making the assessment is placed on the employer, who must ensure that in the case of every work activity—either in progress or yet to be started—an informed judgment is made as to the nature and extent of the risks to health; and that precautions necessary, matched to those risks, are put in place. He is required also to ensure that appropriate channels of communication are provided, so that the requisite information is available to everyone who may need to be kept informed at any time.

The assessment must be sufficient and suitable and this requirement indicates, in an indirect way, who is competent to make the assessment. If the process and risks are simple and obvious, then the amount of expert knowledge required may well be low. On the other hand, where a process is complex or innovative and/or the risks to health are severe or irreversible, then a high level of knowledge will be appropriate. Regulation 2(3) requires that any person who does any work in connection with the duties under this section must have the necessary information, instruction, and training to undertake the task. Thus, the task might be started by junior management within an organization; if no problems arise which are beyond the capability of solution by such staff, no additional help will be required. If, however, in the course of a preliminary assessment, problems are identified which require more expert evaluation then specialists having the appropriate qualifications should be called in. Such specialists may be available already in other branches of the organization concerned, or may be commissioned from external sources.

Guidance from the Health and Safety Executive identifies five main tasks in making an assessment:

1. Gathering information about the work and working practices
2. Evaluating the risks to health
3. Deciding what further precautions may be required
4. Recording the assessment
5. Deciding when the assessment needs to be reviewed

As indicated above, the greater the harm that initial consideration suggests could result from making a wrong judgment about the risks to health, the more care should go into increasing the accuracy of the information—because this is the basis for the decisions to be taken about precautions. Also, if the work is very complex, more effort in detail will be needed because otherwise it can be difficult to be sure that risks have been evaluated correctly. The expertness and degree of detail needed for an assessment will vary from one situation to another, depending on the risks and the complexity of the activities.

Controlling the Hazard

The eventual outcome of an assessment must be reducing to a minimum the risks from exposure to a substance or substances by control of the exposure. It involves consideration of:

(*a*) the likelihood of exposures taking place
(*b*) the peak and average concentrations to which the relevant persons are exposed
(*c*) the length of time over which they are exposed.

If it is not reasonably practicable to eliminate the exposure then Regulation 7 requires that exposure must be controlled adequately. It should be noted that Regulation 7, sub-sections 1 and 3, stipulate that personal protective equipment should be used as the last resort—that is, only when all other feasible measures have been used, so far as practicable, towards achieving adequate control.

A useful check list when controlling hazards is to consider:

1. Elimination of the hazardous substance by,
   (*a*) Substitution by other materials
   (*b*) Changing the process to eliminate use of the substance
2. By plant design,
   (*a*) Total enclosure—under negative pressure where possible
   (*b*) Enclosure so far as practicable, with exhaust ventilation at openings
   (*c*) Local exhaust ventilation
3. Good Housekeeping,
   (*a*) By design of buildings to facilitate ease of cleaning
   (*b*) By keeping areas free from contamination
4. Methods of Work,
   (*a*) Exclusion of personnel
   (*b*) Reduction in periods of exposure
   (*c*) Prohibition of eating, drinking, and smoking
   (*d*) Personal hygiene
5. Provision of information, with adequate training and instruction
6. By supervision

Regulation 8 requires that the control measures, items of personal protective equipment and other related facilities are used or applied properly, and there should be visual checks at appropriate intervals with prompt remedial action where this proves necessary. Employees are required to use the control measures as they are intended to be used, to practice proper hygiene, and to report to management any defects which they may observe.

Maintenance Examination and Test or Control Measures

Regulation 9 requires that measures for control are maintained in good repair and efficient working order.

It requires further that engineering controls should be examined thoroughly and tested; local exhaust ventilation plants should be examined and tested annually (every 14 months)—unless a shorter period is prescribed in Schedule 3 to the Approved Code of Practice. All other plants should be examined and tested at suitable intervals.

Respiratory protective equipment (except disposable respirators) must be examined thoroughly at suitable intervals and tested for efficiency when appropriate. Records must be kept for five years of such examinations and tests.

The Approved Code of Practice requires also that all engineering control measures should have a weekly visual check, where this can be done without undue risk, and that regular preventative servicing should be undertaken.

There is a general requirement that the person who carries out any or all of the above functions should be competent to do so.

The overall object of this Regulation is to ensure that the control measures perform as intended, with any defects which become apparent being remedied as soon as possible. Suitable records must be kept, and in the case of local exhaust ventilation plant specific requirements are set out in paragraph 61 of the Approved Code of Practice.

Respiratory Protective Equipment (RPE)

Paragraphs 62–65 of the Approved Code of Practice set out specific maintenance requirements in respect of respiratory protective equipment (RPE). All items other than one shift disposable respirators should be examined thoroughly and where appropriate tested once every month (more frequently where severe conditions of use exist). For respirators of the half-mask type which are used for short spells, longer periods between maintenance may be allowed—but in no case exceeding three months.

Details of the examination to be carried out and of records to be kept are set out in relevant paragraphs.

Monitoring

Regulation 10 requires that the exposure to substances hazardous to health be monitored by suitable procedures—where this is necessary for the maintenance of adequate controls or to protect the health of employees.

In addition, where a substance or process is specified in Column 1 of Schedule 4 of the Regulations, monitoring must be carried out at the frequency prescribed in Column 2 of that Schedule.

A suitable record or summary containing information on the personal exposure of identifiable employees must be kept for thirty years or, in other cases, for five years.

Whoever carries out the monitoring must be competent to do so and must use suitable occupational hygiene techniques. The presence of airborne contaminants normally will call for the use of personal samplers.

The Approved Code requires monitoring where practicable in the following cases:

(*a*) When failure or deterioration of the control measure could result in a serious effect on health

(*b*) Where measurement is necessary to ensure that an exposure limit or standard is not being exceeded
(*c*) As an additional check on control measures
(*d*) When required under Schedule 4
(*e*) Where recommendations to this effect are made in any relevant technical literature.

Where an assessment shows that monitoring is required it must be carried out every twelve months except in those cases listed in Schedule 4 where more frequent monitoring is required. To simplify the task, monitoring may be carried out on a group basis if individuals in the group are likely to have similar exposures.

Health Surveillance

Regulation 11 requires that the employer shall ensure that employees are under suitable health surveillance where:

1. The employee is exposed to one of the substances and engaged in a process specified in Schedule 5 of the Approved Code, unless that exposure is not significant.
2. The exposure of an employee is such that an identifiable disease or adverse health effect can be related to it, there being a reasonable likelihood that the effect might occur under the particular conditions of work, and when there are valid techniques for detection which can indicate an illness or an effect.

The objectives of health surveillance are:

(*a*) detection of adverse effects in the early stages
(*b*) to assist in evaluation of control measures
(*c*) in collection of data for detection and evaluation of risks to health
(*d*) in the case of micro-organisms, to assess the immunological status of employees.

Health surveillance may consist of:

(*a*) medical supervision by an employment medical adviser or appointed doctor for the special cases set out in paragraph 1 above.
(*b*) medical surveillance by a registered medical practitioner
(*c*) health surveillance by a person with suitable qualifications—as examples, an occupational health nurse, or a responsible person appointed by an employer—who must satisfy himself as to the competence of the individual (such as for the assessment of chrome ulceration).

Suitable surveillance must include always the keeping of a health record for each individual. It may include biological monitoring.

Detailed requirements as to the scope of the appropriate surveillance and the keeping of records are set out in Regulation 11 and paragraphs 77–92 of the Approved Code.

Information, Instruction, and Training

Regulation 12 requires that the employer provides every employee with such information, instruction, and training as will enable the person concerned to know the risks arising from exposure to substances hazardous to health, and the precautions which should be taken. This information should include the results of monitoring, particularly if they should show that the occupational exposure limit has been exceeded, and the results of health surveillance, if required by Regulation 11.

In addition, the information provided should include:

(*a*) the nature and degree of risks which may arise including any factors which may affect that risk, such as smoking
(*b*) the reasons for adopting control measures, and how they should be applied
(*c*) the reasons for, and situations where, protective equipment is necessary
(*d*) the monitoring procedure, and maximum exposure limits where applicable
(*e*) the role of health surveillance and access to records where applicable.

Instruction should be sufficient and suitable so as to ensure that persons who work on the premises:

(*a*) know what they should do and the precautions to be followed
(*b*) know what cleaning, storage, and disposal procedures are required and when and how to carry them out
(*c*) know emergency procedures in case of an accident.

Training must be such as to ensure that all persons at work on the premises can apply and use the methods of control effectively, including wearing personal protective equipment and operating emergency measures if need be.

General

The foregoing gives a general assessment of the impact of the Regulations and Approved Code of Practice in the great majority of premises. For more specific guidance the legislation itself and the detailed guidance provided by the Health and Safety Executive should be consulted. In cases of difficulty, the local office of the Executive should be contacted and the advice of an inspector sought. The most important point to have in mind is where a problem is identified to which a solution cannot be found, the opportunity always should be taken to seek expert advice. Never take a risk where the health of any individual is at stake.

CHAPTER 4

Measurement

J. G. FIRTH

Introduction

The fundamental requirement of any measurement technique is that it should be appropriate for the purpose of the measurement. This means that it should provide the information necessary for the decisions which will be made on the basis of that information. For the Regulations for Control of Substances Hazardous to Health[1] the measurement information is used for two purposes—to assess how much control is necessary, and to ensure that control procedures are adequate. Thus, basically the measurement process is part of the assessment procedures under the Regulations.

The primary question asked of any measurement is 'What do the results mean?' What are the norms against which they must be judged? Essentially, this is to ask when is a situation identified by a measurement 'satisfactory' and when is it 'not satisfactory'? Quite rightly, occupational hygienists will argue that when making an assessment at a particular place of work considerable professional judgment should be brought to bear—but in very many instances the final view of whether a situation is 'satisfactory' or not rests on whether the concentration of a given substance is above or below the relevant exposure limit.

The occupational exposure limits for particular chemicals used in Britain are published each year by the Health and Safety Executive in Guidance Note EH 40.[2] The processes by which the limits are established are described elsewhere in this book but essentially in Britain two types of exposure limits apply.

Of the two types, the most important are the Maximum Exposure Limits ('MELs'), which are for substances which can be very harmful (*e.g.*, carcinogenic) or where experience shows that the costs of reducing occupational exposure to a level where harm to health is negligible are too great—so that for the substance concerned a limit has to be set above that at which no effect is anticipated. As a corollary to this, the law then requires that exposure to such a substance should be reduced as far below the maximum exposure limit as is 'reasonably practicable'. This means in effect that the enforcing authority can deal with particular situations on an individual basis. Currently, MELs are set only for about thirty substances, but the nature of these substances and the stipulation about reducing exposure means that in many instances their measurement is a continuing requirement. Since changes in the concentrations of these substances in the workplace will give rise to important decisions it is necessary to have validated methods for their measurement.

The second type of exposure limit in use in Britain is the Occupational Exposure Standard ('OES'). Such standards are applied to some 500–600 chemical substances and are the concentrations below which the best evidence available indicates that there are no adverse effects on health. However, some of them were based on limited or incomplete evidence so that naturally, as more toxicological and other data become available, they are subject to change. The essential feature of the OES for a substance is that it is the level below which the Regulations require an employer to take little or no further action. In essence therefore, if measurements are made and the amounts of a particular substance are found to be below the occupational exposure standard then there is little need for the employer to take further measurements (or, indeed, any further action).

Measurements for a particular substance at a particular time tell only part of the story and it should be remembered that the levels may change with time, thus indicating, even though concentrations still may be well below the standard, whether or not a situation is deteriorating. There may be changes also associated with changes in a process and these can indicate that some improvement in control procedures is needed.

Examination of Guidance Note EH 40 shows that most of the exposure limits quoted relate to concentrations in air—that is, it is assumed that the primary route of access to the body will be via inhalation. However, this of course is not the only route by which substances can enter the human body. Many also can penetrate the skin, and they are indicated in EH 40 by an (S) or 'skin' notation: it is possible too that if personal hygiene practices are inadequate substances will be taken by mouth. The only sure way of measuring exposure to substances entering by routes other than inhalation is by applying methods which measure amounts taken into the body. Measurements obtained in this way generally are said to be obtained by 'biological monitoring methods'. Usually such methods measure the amount of a substance or of one or more of its metabolites in one or other of the two accessible body fluids—blood or urine. In the case of a substance which is relatively volatile there may be equilibrium between the concentration in air in the lungs and in the blood, and in such situations the monitoring of breath is an acceptable approach. Increasingly too, saliva is regarded as a material that can be obtained by non-intrusive sampling and the concentration of a substance in saliva can be used as an indication of the amount present in the body.

'Biological monitoring' is mentioned specifically only under Section 11 of the Regulations but it is accepted that exposures may be quantified by means of any suitable technique; there is no reason to think that 'biological monitoring' might not be acceptable for purposes of occupational hygiene.

It appears likely that biological monitoring will be of increasing importance in the future as a means of assessing exposure under the Regulations. At the present time, only lead in Britain[3] has exposure limits set in terms of concentration levels of the substance or its metabolites in one or other of the body fluids. However, such levels are being set in the United States[4] and Germany[5] and it may be anticipated that moves in this country will be increasingly in this direction in the future.

Requirements for Measurement

Almost all the occupational exposure limits published in Britain relate to levels of the substances concerned in air. The regulatory philosophy is that they should be related to the concentrations in the air in the breathing zones of the employees concerned, in an attempt to estimate closely the quantities of the substances that could be absorbed by inhalation during the period of work.[6] For substances representing chronic health hazards, almost always the working period is taken to be eight hours—that is, the average working shift. One exception to this is vinyl chloride,[7] for which the exposure limit is in terms of a twelve month period. (In this instance, conversion tables are provided to assist in relating short-term measurements to one-year exposure.) Its inclusion among the limits for this country is a consequence of a limit being set in this way originally by the European Commission; it recognizes that the hazard for the monomer is a consequence of long-term exposure to low levels. However, this more complicated, though sensible, procedure was not followed in Britain for any other substance with toxicology similar to that of vinyl chloride.

For a number of substances, short-term exposure limits also are set. Such limits represent an attempt to recognize the hazards which may be presented by exposure of individuals to high levels of certain substances for short periods of time—and to limit such exposures accordingly. Such limits are also a recognition that in many processes the actual periods of exposure to the substances concerned may be over brief periods of time and that it is important therefore to set maxima for such circumstances. Most frequently the period for short-term limits is ten minutes, which was derived originally from an old idea that such a duration was necessary at normal rates of sampling before a satisfactory sample could be obtained. Improvements in monitoring instruments and in the increased sensitivity in analysis open the possibility of reliable measurement with samples taken over shorter periods.

It will be seen from a brief examination that most of the exposure limits are in parts per million by volume. Thus the three most important factors which determine the requirements for the type of equipment and the techniques for sampling and analysis may be summarized as:

(i) the need to monitor personal exposure
(ii) the value of the concentration limit in air
(iii) the period of time over which the limit is permitted.

Strategies for Measurement

The procedures to be followed in assessing risk from exposure to hazardous substances while at work are set out in Guidance Note EH 42[8]. They include the identification of the substances, of the processes in which they are used, and the procedures for their storage and use. On occasions when it is established that there is at least a distinct possibility of exposure of human beings then it may become necessary to ascertain by measurement the extent of the exposure. It will be found in many situations, even under the worst circumstances at the place of work, that the concentrations of the substances are well below the limits given. However, there may be particular areas of a plant, or particular processes, where risks seem to be more significant than elsewhere. It can be helpful, therefore, to make a fairly rapid and inexpensive 'screening' of the entire unit to find such areas, and then to employ the more precise measuring techniques as is seen to be most appropriate.

Guidance Note EH 42 identifies three broad groups of strategies, each at an increasing level of sophistication. The overall aim of this approach is to obtain at reasonable cost information of the relevance and quality required.

The first level strategy is intended to give an impression of a situation, from which it can be decided whether a risk might exist. At this level, sampling equipment and measuring techniques are relatively simple and very often the strategy will be to take the 'worst case' and use this as an index for overall risk. The techniques used might involve the 'dust lamp' for identifying maximum emissions, and spot readings with devices such as colorimetric tubes. Often these will be helpful in acquiring a general picture before more precise measurements are undertaken. It may be that the pattern of exposure is very variable, depending upon the work programme and the actual distribution of operators in the unit, and in such circumstances it can be worthwhile to divide a total population into groups, each of which is as homogeneous as possible so far as the types of work undertaken and the likely periods of exposure are concerned. The periods of high or peak exposure should be identified and sampling conducted during these phases. Emphasis should be placed on measurement of personal exposure, although not necesarily over the whole of the period of time involved; it is permissible also to employ static or background sampling if this will help to make the picture more complete. When designing a programme of sampling in order to verify a situation, care should be taken to obtain sufficient samples among members of the group or groups of operators thought to undergo the highest risks.

The first level or 'screening' strategy should indicate whether or not significant risk might exist. If it does, the advice would be to move to the second level of strategy for sampling and measurement, at which the object is to obtain accurate measurements which can be related to the limits given in Guidance Note EH 40. Whenever possible, sampling should cover the entire period of exposure of an individual—taken either in one or in two or more consecutive procedures. (The taking of consecutive samples can be advantageous in that it might help in identifying changes in conditions in the course of a shift, and when the maxima occur.) Often it is sensible to employ a 'stratified' approach to sampling, in which effort is concentrated upon the periods of maximum risk. Compliance or otherwise with both the long-term and short-term limits then can be assessed.

When different groups of employees undertake similar tasks under the same conditions the sampling can be on the basis of one group. However, if there should be considerable variation in the work on different shifts for each group of people (differences between the day shift and the night shift), then the groups concerned should be assessed separately. It may not be necessary to sample for each individual in the group but a sufficient number of members should be selected on a random basis to give an evaluation for the group as a whole.

Whatever strategy is used should enable the exposure of individuals or groups to be assessed in relation to the relevant limit or limits. In most instances the eight hour time-weighted average will be the standard appropriate for comparison, and the sampling and analysis should provide data for such a comparison. This can mean in many cases that sampling should be carried out throughout the shift or working day. If there should be considerable variation in the pattern of work from day to day it may be necessary to sample on a sufficient number of different days to cover all variations.

For substances which have MELs it is appropriate almost always to apply a second level strategy for measurement. Unless it can be shown that occasional results above an MEL are without real significance the Regulations require immediate action to gain control and to reduce exposure below the limit. Such determinations call for more than a straightforward screening or first level appraisal.

The results from the survey should make it possible to answer the following questions:

(i) is action necessary immediately to reduce or to eliminate exposure?
(ii) is action necessary immediately to re-establish control?
(iii) is a programme of improvements required?
(iv) is a more detailed survey required?
(v) should routine monitoring be continued or implemented?

Occasionally, a sampling programme with a high degree of sophistication will be wanted. (This would be regarded as the 'third level' of strategy.) An instance of when this might happen would be where all reasonable control measures were applied and yet exposure remained close to the relevant limit. In such a case a routine programme of monitoring with the ability to record quite small increases or decreases in concentration might be necessary and unless data were collected in accordance with a rigorous protocol, coupled with assessment by means of suitable statistical techniques, it might be that such small changes were not identified.

Biological Monitoring

Biological monitoring[9] is relevant particularly for substances which can be absorbed through the skin as well as taken in by inhalation. It is a useful technique too where there is dependence for control of exposure on wearing personal protective equipment—so that the measurement of airborne levels of the substance or substances is not appropriate. A further advantage is that it can reflect the different rates of absorption of a substance by different individuals—arising from differences in their work practices and in their rates of metabolism. A measurement made in this way is personal to the employee and might well encompass other features of his (or her) habits or health; hence it is important that rights be safeguarded and that the way in which results are presented ensures a suitable degree of confidentiality so far as the individuals are concerned.

The rate of absorption of any substance depends on a number of factors, among which are:

(i) differences in height and in other aspects of physique
(ii) general condition of respiratory system
(iii) variations in the propensity to absorb through the skin
(iv) variations in the propensity to retain materials soluble in fat.

The degree of obesity or otherwise of a person can affect the concentrations of the substance found in the different organs or body fluids[10] and a weakness of taking airborne measurements is that they will not show such variations between individuals. The interplay between biological factors, differences in behaviour and in the conditions of work can result in quite considerable variation from person to person in the extent of absorption.

Different substances will be absorbed through the skin, metabolized, and excreted at different rates. It is very important to keep in mind that many substances are metabolized, so that over time there is a continuing reduction in the levels of substance and metabolites present. (A metabolite is a product of decomposition or modification—in general a form of the substance more soluble in water and derived either in the kidneys for excretion in urine or by the liver for passage through the intestine.) The rate at which a substance or metabolite disappears from blood, breath, or urine is expressed as a 'half-life', or the time taken for the concentration in the medium concerned to fall to half that prevailing at the end of exposure. Half-lives may be measured in minutes, hours, or days. The strategy for sampling in blood, urine, or exhaled air should be related to the half-life of the substance to be measured. In general it will not be convenient to apply biological monitoring to

TABLE III *Biological half-lives of certain substances*

Substance	*Half-life* (hours)
Benzene	3
Carbon disulphide	2
Dichloromethane	0.5
Dieldrin	3500
Lindane	150
Mandelic acid (metabolite of styrene)	5
Mercury	90
Pentachlorophenol	400
Perchloroethylene	15
Polychlorinated biphenyls	50 000
Styrene	0.7
Toluene	5
Trichloroethane	6

substances with half-lives of less than two hours. Substances which have half-lives of between two and ten hours can be sampled most suitably at the end of a shift or on the following morning. Substances with half-lives of between 10 and 100 hours can be sampled at the end of a shift or at the end of the working week. In cases of half-lives greater than 100 hours, sampling can be random. Table III shows the half-lives of several substances and metabolites for which this form of monitoring is undertaken.

While on the subject of biological monitoring it is worth mentioning that such samples always should be taken under appropriate conditions, by personnel who are qualified and experienced in such work.

Methods of Measurement

(i) Screening Techniques for Gases and Vapours

A wide range of devices is available for easy screening and rapid measurement of airborne gases and vapours. One of the simplest and most versatile uses the colorimetric tube—a transparent or translucent tube in which one or more active chemicals may be supported by an absorbent such as silica gel.[11] A hand-operated pump is employed to draw a known volume of atmosphere through the tube. If the sample taken contains the relevant gas or vapour there is a reaction with the chemical reagent in the tube, to give a coloured stain. In some instances the concentration of the gas or vapour may be estimated from the depth of the colour of the stain, but more usually this is done from the length of the stain within the tube.

Such detector tubes can be used in measurements for a wide range of compounds, both organic and inorganic; their precision ranges from 5% to 30%, depending largely upon the reaction principle employed for a particular substance.

Certain minimum requirements for detector tubes were set out in British Standard 5343: 1976,[12] including their marking, shelf-life, and accuracy, manufacturer's instructions, and the calibration and use of aspirating pumps. Briefly, the tubes should remain within specification under all humidity conditions over the temperature range 5°C to 35 °C, and after storage for at least one year at temperatures up to 30 °C. The pump should have an accuracy of not more than 10% above or below the volume of air sampled. Where such requirements cannot be met, manufacturers should state the limitations. The Standard requires also that details be given of the chemical reaction in the tube, with a clear statement of the possibilities of interference—especially when these could give rise to low results.

Unfortunately, at the time of writing it is understood that none of the tubes provided in this country has been certified to the British Standard. Various organizations have made evaluations for their own purposes and these are believed to suggest that in general terms the detector tubes from the three main suppliers in Britain either meet or come close to the requirements of the Standard.

There is an extensive and growing range of instruments for making rapid spot measurements of gases and vapours in atmosphere. In most cases these represent outlays of capital much greater than for the colorimetric tubes, and this in turn suggests that a purchaser should need to make much larger number of measurements within a given period of time. The requirements for maintenance of these instruments should be followed carefully in order to ensure their reliability. At the present time there is no British Standard stipulating minimum requirements for performance, although such a document is in preparation. Until the Standard is available one would be well-advised to follow closely the manufacturers' instructions and to question manufacturers and suppliers, especially with regard to the possibility of interference by other gases and vapours.

The critical part of any instrument of this type is the transducer, which converts the measurement of concentration of gas or vapour into some form of electrical signal. Factors such as performance, range of operation, reliability, and the effects of gases interfering, depend almost entirely on the characteristics of the transducer. Modern instruments use a wide range of different transducers[13] but in general in small portable measuring devices the types found are electrochemical cells and simple optical or spectroscopic systems.

Frequently, instruments which are based on electrochemical cells are sold to measure one, two, or three gases or vapours—these almost always being simple, from a chemical point of view, and reactive (such as carbon monoxide or hydrogen sulphide). Very often the reaction on which the measurement is based is oxidation, which means that chemically similar gases sometimes can give similar signals—so when preparing to take measurements in mixtures of gases containing similar components it is advisable to consult the manufacturer's handbook with special care. It is possible also that some gases may suppress a signal from a vapour it is desired to measure, and again the manufacturer's advice should be followed.

Spectroscopic instruments are likely to have a greater specificity to individual substances but it seems that there are difficulties in constructing a highly-specific spectroscopic system at a cost low enough for the cheap portable instruments. With these systems also, whenever it is desired to measure similar gases or vapours, advice should be obtained from the manufacturer.

The Health and Safety Executive maintains a large computer record of instruments for the measurement of gases and vapours and information from this source is available to enquirers.[14]

(ii) Instruments for Measuring Dusts and Fibres

For the most part, techniques which are used for rapid measurement of dusts and fibres in air are based on physical rather than chemical processes. Users may well find such techniques easier to comprehend, and generally the limitations and possibilities for interference will be more evident.

The so-called 'dust lamp'[15] is the easiest method of screening for dust. Basically it is a powerful source of white light which can be used to illuminate dust in a working area by forward scattering (the 'Tyndall effect'). The method makes it possible to see very fine particles of airborne dust which are not visible under normal lighting. It does not provide quantitative measurement of the concentration of dust but assists by indicating the sources of emission.

In recent years several types of small readily-portable instruments have been made available for the monitoring of dust on a continuing basis.[16] They apply light scattering techniques and have no chemical specificity. However, they can be calibrated for particular types of dusts and the semi-quantitative signals generated can enable easy and rapid estimation of variations in the concentrations of the dust within the unit being studied. These instruments show readily whether or not the concentration of dust is near to the relevant exposure limit.

When using instruments of this nature it is important to bear in mind that they are not specific; the measuring principle relies on the size of the particles of dust and if after calibration the average particle size or chemical composition of dust at a place of work changes, readings obtained under the changed conditions will be ambiguous.

Quantitative Measurement of Airborne Contaminants

A range of instruments is available for quantitative measurement of airborne contaminants but in general the application is limited to the chemically simple gases and vapours. In any case, the range of substances used by industry is so wide that the scope for instrumental measurement inevitably would be somewhat restricted. Equipment available covers toxic gases such as ammonia, carbon monoxide, chlorine, hydrogen sulphide, nitrogen oxides, and the simpler hydrocarbons. These are some of the single gases encountered most widely in industry and there is wide application for portable instruments both for making surveys for them at places of work and for regular monitoring. There are also, for routine monitoring, many fixed multi-headed systems of considerable complexity. Often the type of measurement technique employed is based on a laboratory procedure such as gas chromatography: the limitations of the instruments and systems would be those of the techniques employed.

To an increasing degree, types of instrument usually associated with the laboratory—particularly infra-red spectroscopy—are being constructed in the form of portable units which can be used to take measurements in the field.[17] Some excellent devices are available for atmospheric monitoring and they have the versatility of infra-red spectrometers in that they can be used for a large number of different gases with a considerable degree of specificity.

Portable versions of gas chromatographs also are being developed; they have the inherent flexibility of range and application associated with the technique and are likely to be used to a larger extent in future.

These developments notwithstanding, quantitative measurement for the majority of gases and vapours depends still on the application of chemical analytical science. This means in practice that atmosphere is sampled for given periods of time through some form of filtration system which removes the substances of interest for quantitative analysis later at the laboratory. The process of measurement thus falls into two stages—sampling, followed by analysis and calculation of results.

(i) Samplers for Gases and Vapours

Traditionally, the sampling of atmosphere for gases and vapours would be by means of a sampling pump[18] which would draw through a filtration material (that is, an absorbent) over a fixed period of time a known volume of air. A typical filtration material would be activated charcoal, which would be contained in a tube or some other suitable vessel. Other absorbents used, in tubes of various sizes, have included: alumina, molecular sieves, and silica gel. The amount of substance collected was determined by suitable analytical techniques and by calculation (from the period of time of sampling and the rate of flow of the air) the average concentration in air at the time of sampling could be recorded.

Among the types of sampling device employed have been:[19]

(*a*) tubes containing suitable solid absorbents, as described above

(*b*) bubblers, containing a suitable absorbent liquid—or, more often, a reagent dissolved in a liquid

(*c*) impingers, in which a jet orifice was used to direct the gas sampled to the surface of the liquid.

There are some obvious disadvantages in using liquids as absorbents and development has been directed towards the use of solid materials for this

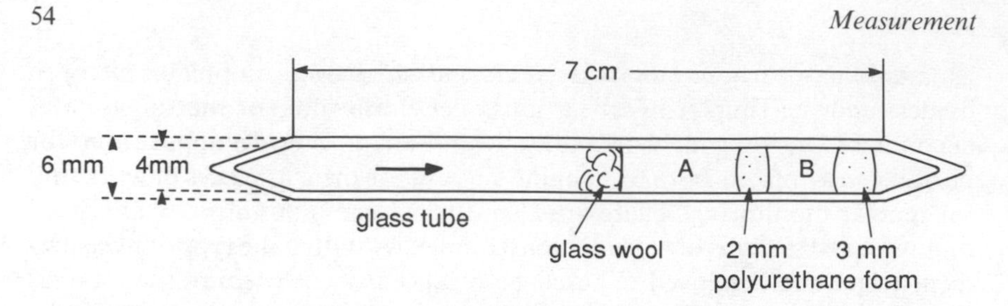

FIGURE 1 *'NIOSH' vapour adsorption tube ('A' and 'B' are beds of charcoal of 20/40 mesh,* 100 mg *and* 50 mg*, respectively.)*

purpose. There would be benefits also from the availability of solid absorbent in a standard form of tube, which could be made in fairly large numbers with uniform characteristics, and the training of operators in the use of such tubes is straightforward. There are good reasons also for trying to reduce the size and weight of the sampling pump, if this has to be carried or worn by an employee in the course of normal duties. Some pumps are bulky and heavy, and hence inconvenient to use, so there have been parallel efforts to make both sampling devices and pumps as small as practicable.

The first standardized miniature absorbent tube to be used in large numbers was developed by the U.S. National Institute of Occupational Safety and Health and is known as the 'NIOSH' tube.[20] Essentially, it comprises a small glass tube containing two beds of charcoal (see Figure 1). Tubes of this type are sealed until required for use, when the tips are broken off in preparation for sampling. Air is drawn through the first bed of charcoal and the second is used to detect any breakthrough of a substance from the first bed.

The main advantages of these tubes are their small size and ease of use. The charcoal is in standard form and is a good absorbent for a wide range of organic vapours. Since the tubes are small, the air pumps employed with them also can be small and easier to carry or to wear.

After sampling, the tubes are closed by means of plastic end caps and taken to the laboratory for analysis. At the laboratory, each of them is opened and the charcoal removed carefully into glass vials; a solvent which will take up the substance absorbed is added to the contents of each vial, made up to volume and injected into the appropriate analytical instrument.

The deficiencies of this method of sampling are that the extraction with solvent involves dilution of the absorbed material and can be time-consuming. Often, recovery of the absorbed material is not complete, and for many substances it is necessary to prepare calibration curves[21] for recovery. These features have the effect of limiting the sensitivity of measurement. Also, a single sampling tube can be used only once. To some extent deficiencies of this nature have been met by the introduction of a wider range of absorbents, including proprietary materials such as Tenax and Chromosorb, and others that are employed frequently in packings for chromatography.

Exposure limits are always under review and reductions are made fairly frequently. This in turn gives rise to requirements for greater sensitivity of measurement and has led to the development of samplers in which material absorbed can be desorbed directly by heating, into the measuring system. Such devices make use of the fact that most systems measure in the vapour phase. Absorbent is contained in metal tubes which can be flash heated and flushed with inert gas. The desorbed vapour is carried away in the flushing gas and either introduced into the analyser or stored in a reservoir. Under this technique, preparation of the sample is eliminated, there can be no effects from any impurities in solvents used for desorption, and sensitivity is increased greatly because the measuring system analyses the whole sample. In general the recovery of the sample is more complete and the tubes are again clean and can be used again after each analysis.

The application of thermal desorption made possible the most important development in recent years in techniques of sampling for gases and vapours—the so-called 'diffusive sampler'.[22] Such samplers do not require pumps, the rate of sampling being governed by a layer of air contained either in a small part of the sampling tube or in a porous polymer. The air layer separates the ambient atmosphere from the absorbent and limits the rate of diffusion of substances from atmosphere to absorbent. In practice the rate at which sampling is carried out by such devices is much lower than with pump-operated samplers—typically one millilitre per minute, compared with, say, 100 millilitres per minute—but the greater sensitivity of thermal desorption more than overcomes this disadvantage. Figure 2 shows one of the best-known types of these samplers.[23]

Thermal desorption can be used in the laboratory for large numbers of samples in automated systems; the reduction in manual intervention improves considerably both accuracy and reliability.

Diffusive samplers are available in several geometrical forms. The two best-known are the tube (as described above) and the 'badge'.[24] In the latter the ratio of the area of cross-section of the diffusive path to its length is much higher than in a sampler of the tube type—which means that the rate of sampling is much increased (by one or two orders of magnitude). As shown

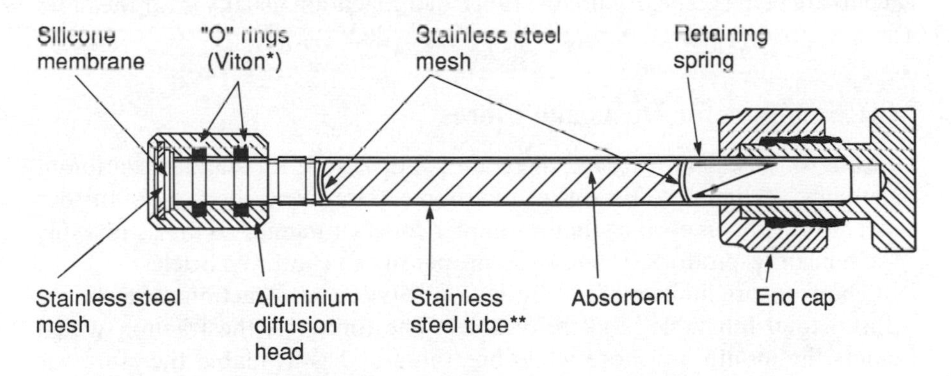

* Registered trademark for fluoroelastomer
** Length 3½ inches; outside diameter ¼ inch

FIGURE 2 *Thermal Desorption Tube*

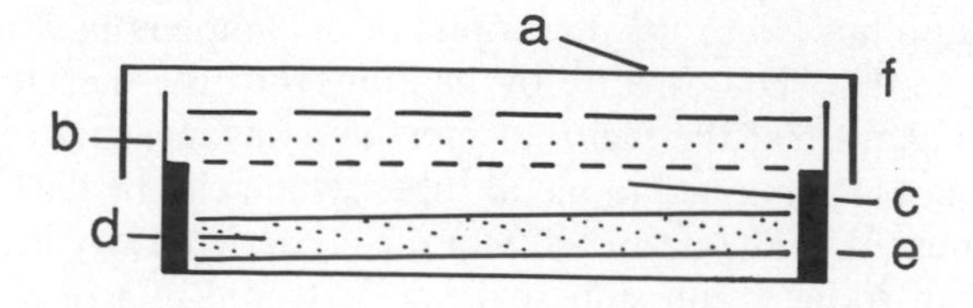

FIGURE 3 *Diffusive Sampler:* (a) *Removable cap,* (b) *membrane (optional),* (c) *diffusion layer or gap,* (d) *sorbent,* (e) *spacer,* (f) *windshield.*

in Figure 3, the absorbent behind the diffusive barrier can be a solid but it is possible also to use liquid absorbents which have the required degree of specificity.[25] Such devices offer higher sensitivity but to some extent this is offset by the need to remove absorbed vapour from the medium by means of solvents or other methods involving manual manipulation. Very often this means that a sample has to be diluted, and whenever an analysis requires a greater number of manual stages there is risk of reduced reliability in the results.

In the field of diffusive sampling a further important development was the introduction of a range of colorimetric sampling devices: these make measurements on the sampler at the site, so there is no need for analysis at a laboratory. In one elegant system, gas passes through the diffusion layer into a tube of colour reagent, the length of the stain developing over time in the tube providing a measure of the average concentration of a substance in atmosphere in the sampling period.[26] A variety of such samplers is available and they make possible ready determination of time-weighted averages over periods up to eight hours. They are simple, convenient to use, and call for only a limited amount of training of operators.

A parallel development of the badge-type device has been the use of impregnated paper systems for absorption.[27] The concentration of a vapour in the air is shown by the intensity of stain developed on the paper during the period of sampling. Devices of this nature may be less accurate than samplers of the tube type but their development is proceeding and improvements are being made. Again, the range of application usually is for the more simple and reactive gases, but is being expanded.

(ii) Samplers for Dusts and Fibres

Just as for gases and vapours there are good reasons for standardization in sampling techniques for dusts and fibres. However, there is a further incentive—it is important that a sampler for dust mimics so far as possible the human respiratory system in its propensity to capture particles.

The exposure limits make a distinction between two fractions of airborne dust—'total inhalable' and 'respirable'. The former is the fraction which enters the mouth and nose while breathing and is available therefore for deposition anywhere in the respiratory tract. 'Respirable' dust is the fraction which penetrates to the gas exchange region of the lung: the convention applied is that recommended in 1952 by the British Medical Research

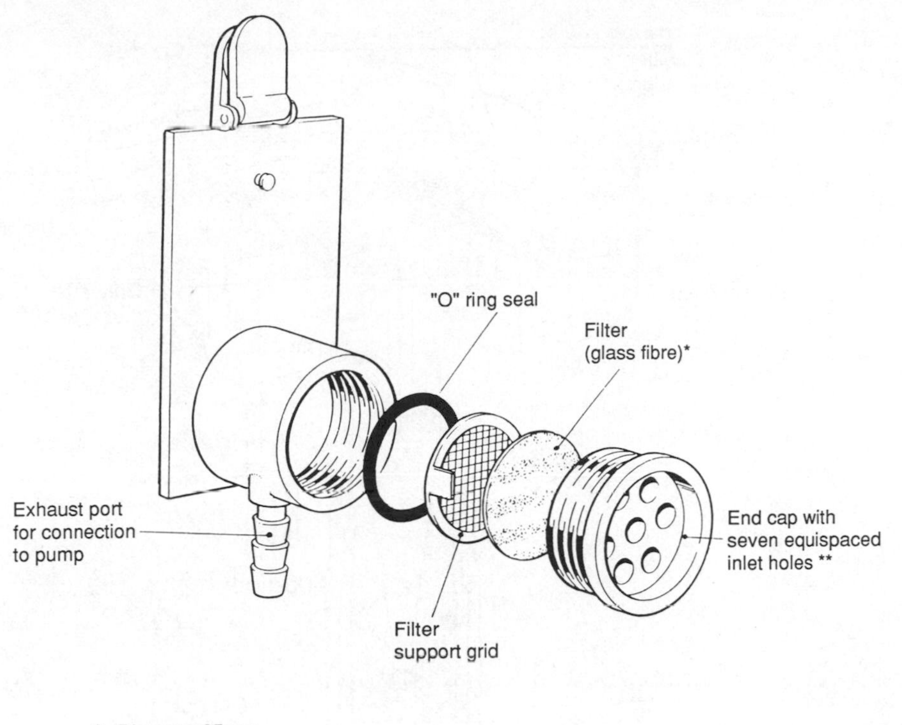

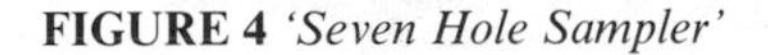
FIGURE 4 *'Seven Hole Sampler'*

Council and adopted in 1959 by the Pneumoconiosis Conference at Johannesburg.[28]

Important types of materials among inhalable dusts are those which are soluble in body fluids—in other words, any such material which enters the resiratory tract can appear eventually elsewhere (in other organs) as a result of transport in the blood. Materials of this type would include dusts and fumes of metal such as lead and certain cadmium compounds.

Material not dissolved in body fluids but of particle size such as to be respirable can remain in the gas exchange region of the lungs and cause difficulties: some 'traditional' occupational diseases have their origin in materials of this nature—such as pneumoconiosis caused by coal dust, and silicosis from silica.

The characterization of samplers by their ability to capture particles of different sizes can be very time consuming and it is important that standardized samplers by developed and used as extensively as possible. The types recommended in Britain for the two types of dust are described in the Health and Safety Executive publication 'Methods for Determination of Hazardous Substances' No. 14.[29] That for 'total inhalable' is a modified form of a sampling head used by the Atomic Energy Authority and known as the 'seven hole sampler' (see Figure 4).

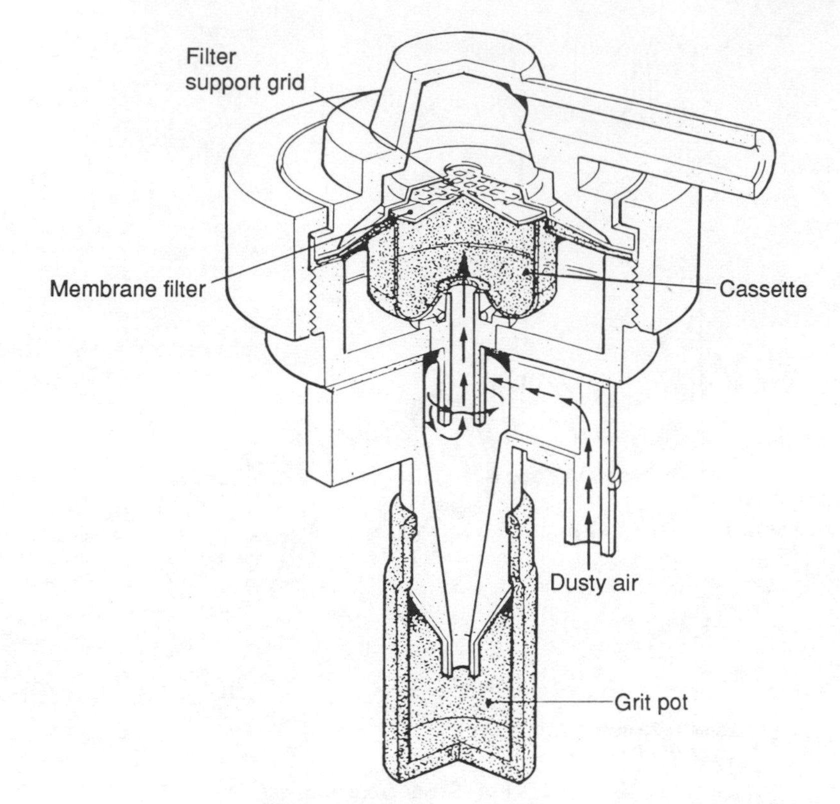

FIGURE 5 *Cyclone Pre-Selector*

The sampling of the 'respirable fraction' requires a more sophisticated system for selecting particle size and in personal sampling a cyclone pre-selector is the most usual (Figure 5).

Such devices require a personal sampling pump, and in general operate at a rate of flow of about two litres per minute. Their ability to separate particles depends critically on the rate of flow and when using them it is essential to follow closely all the recommendations about operating conditions.

In both types of sampler the selection of the filter is very important. Very often, analysis of the sample is by gravimetry only. Filters made from certain materials—such as cellulose nitrate—can absorb moisture and unless due care is taken this may increase the weight of the filter by more even than the material sampled.[30] Filters which do not absorb moisture so readily—such as vinyls—can acquire static electric charges which not only interfere with the weighing but also will draw down further particles of dust after sampling has been completed. The material of the filter must be suitable for the purpose in hand, and the particular characteristics and requirements taken into account fully.

When sampling for asbestos fibres, standard equipment and procedures

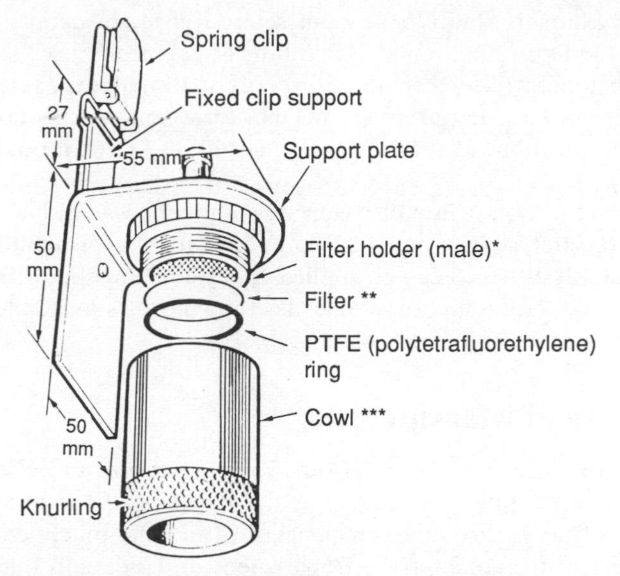

* Such as for Gelman Sciences type 1107
** Diameter 25 mm
*** Length 47 mm; internal diameter 22 mm; outside diameter 32 mm. Recessed and threaded to match filter holder (male)

FIGURE 6 *Open Head Sampler*

should be used. The device recommended (although similar samplers can be employed also) is described in 'Methods for Determination of Hazardous Substances' No. 39/2.[31] Essentially, such devices are open-head samplers as shown in Figure 6.

(iii) Analytical Methods

After sampling, the whole range of laboratory analytical methods is available for determining the amounts of material in or on the filter media.[20,32–34] Quite naturally, the techniques used most commonly for gases and vapours are: infra-red spectroscopy, gas chromatography, and, for more complex substances, mass spectrometry. Increasingly, for organic substances which are not volatile, high performance liquid chromatography is used.

When chemical analysis of dusts and fibres is required, techniques such as atomic absorption and *X*-ray fluorescence spectroscopy may be employed; however (and as usually is the case), when an exposure limit for a substance is expressed as mass per unit volume and it is clear that only that substance occurs in the area sampled, a simple gravimetric determination may be all that is necessary.[35]

The range of analytical techniques applicable in different circumstances is too wide to be described here but methods for substances for which maximum exposure limits have been set are given in detail in the series 'Methods for Determination of Hazardous Substances'.[34]

For substances for which there are occupational exposure standards, methods are covered in part in the MDHS series and by the more extensive range of publications of the National Institute of Occupational Safety and Health.

In this connection, a useful recent development was publication by the Health and Safety Executive of a computer database of methods for sampling and analysis which can be applied to all substances that are subject to maximum exposure limits or standards: the record is available on 'floppy disc' and will be brought up-to-date each year.[36]

Reliability of Measurement

The value of these investigations for the Regulations and other purposes depends upon the quality of sampling and on the subsequent analysis. The reliability of analytical measurement is attracting much comment and formed part of the content of a Department of Trade and Industry White Paper published under the title 'Measuring up to the Competition'.[37]

Where measurements are carried out by properly-trained personnel, reliable results depend upon the following:

(*a*) the availability or otherwise of standard reference materials. These enable proper calibration of instruments to be used in the field and at the laboratory.

(*b*) use of a validated method of test. For instruments, this would imply prior testing against an appropriate standard. For analytical methods, it would mean that they have been evaluated against a given protocol.

(*c*) proper quality control procedures. For field instruments, this means they are maintained properly and checked frequently using standard atmospheres. For laboratory methods, it means regular use of 'blank', 'check' and/or 'spiked' samples, together with participation in relevant schemes for comparing results obtained in similar work by different laboratories.

Reference Materials

Many of the suppliers of instruments to be used on site supply also aerosol cans of known concentrations of the gases or vapours to be measured, and these may be used for setting up the equipment correctly. However, although such cans generally give reliable results it is not unknown for the concentration of the contents to be markedly different from that indicated: with this in mind, it is advisable in practice to check the calibration of an instrument using two independent suppliers of reference materials.

For purposes of laboratory analysis, the availability of reference materials covering the types of samples taken in occupational hygiene is very limited. A programme for the production of such materials was established by the

European Bureau of Community Reference but it requires two or three rounds of comparisons of analytical results obtained in ten or more laboratories in different countries and this is bound to take time. At the date of writing this, the range of such materials available is small—only aromatic hydrocarbons (benzene, toluene, and xylene) are available on Tenax. A further programme is under way to produce:

on charcoal: benzene, toluene, and xylene
on impregnated charcoal: vinyl chloride monomer
on Tenax: chloroalkanes, esters, and ketones.

Bearing in mind the wide range of substances met with in practice in work of this nature it seems unlikely that a significant range of reference materials can be expected in the foreseeable future.

Methods of Measurement

Protocols or procedures for the validation of diffusive samplers and of pumped samplers for gases and vapours are described, respectively, in 'Methods for Determination of Hazardous Substances' No. 27[38] and No. 54.[39] Whenever possible, samplers for this purpose should be validated against these protocols or similar evidence of the performance they offer should be available. The protocols do not stipulate requirements but it is accepted that the accuracy overall of a measurement process (that is, taking account of variability in both sampling and analysis) should be such that 95% of measurements are within 30% above or below that stated.

Samplers suitable for dusts and fibres are described in MDHS No. 14[29] and No. 39/2.[31] Standard samplers such as these have the performance characteristics required.

The validation of analytical methods is more complicated. Protocols for this purpose were described by NIOSH.[40] However, full validation against a protocol can be time-consuming and in practice there is a tendency to limit this to methods where the results obtained depend very much on the procedure employed. An extreme example would be the measurement of asbestos by phase-contrast optical microscopy, where the result of fibre counting is dominated by the rules followed. In this instance a standardization European method was developed, and this is the one described in MDHS No. 39/2.[31]

The Community Directive on exposure limits seeks to establish limits applicable throughout Europe and thus draws attention to work on the development of European Standards for measurement in this area.[41] The European Centre for Standardization ('CEN') is responsible for preparing these documents, more particularly its Technical Committee 137. This committee will develop Standards specifying performance required for methods of measurement, together with protocols for their validation, but it is not the intention to prepare a list of standard analytical methods. However, it is expected that most of the analytical methods published already in the MDHS series and by NIOSH will meet the requirement overall of approximately plus or minus ten per cent as a sum of precision and bias.

Control of Quality

'Round robin' trials and other inter-laboratory comparisons have shown that even where standard or other widely-used methods of analysis are being employed, with what are believed to be proper calibration procedures, there still can be a wide variation in the results obtained in different laboratories.[42] Such observations have drawn attention to requirements for the control of quality of results both within laboratories and through comparisons of performance in inter-laboratory schemes.

An MDHS is available giving advice on control procedures to be employed within a laboratory.[43] Such procedures are applied more readily when large numbers of samples are being analysed for the same or similar parameters. However, many laboratories often are analysing a wide range of substances and the number of samples of each substance is relatively small. In such situations it is more difficult to apply quality control procedures, although it is reasonable to suppose that a laboratory that cannot cope adequately with suitable control measures when samples are analysed in large numbers would not be likely to be better with smaller numbers of the type described.

Inter-laboratory schemes require the circulation of similar samples within as large a group of laboratories as possible and comparison of the analytical results from one laboratory with the mean of those from all the rest. The Health and Safety Executive operates a scheme of this nature for asbestos fibre counting, under the title 'Regular Inter-Laboratory Counting Exchanges' ('RICE'),[42] and there is a similar scheme for typical occupational hygiene samples (such as metals on filters and vapours on charcoal), with the name 'Workplace Analysis Scheme for Proficiency' ('WASP').[44]

In a scheme of the WASP type, the performance of each laboratory taking part is measured by an index ('Performance Index', or 'PI') for each round of samples. Preparing the index for a given laboratory involves calculating the sum of the squared deviations from 1.00 of the standardized results divided by four. (This is equivalent to the variance of the standardized results about one.) In order to avoid small figures, the basic index is multiplied by 10 000. The lower the PI for a laboratory the better its performance. In summarizing the outcome of a round, the indices for all the participants are grouped in three categories, as follows:

1. good performance
2. adequate performance
3. unacceptable performance.

As it happened, in the first year of operation of the scheme, almost half the laboratories participating were in Category 3.

Experience gained in running the RICE and WASP programmes provided a basis for a protocol for inter-laboratory 'quality assurance' schemes. The protocol was agreed by an *ad hoc* committee of organizations operating such systems in the chemical and clinical chemical fields and is hoped to lead ultimately to an international protocol on this subject.

Interpretation of Results

The underlying purpose of the equipment and procedures described here is to ensure that the measurement obtained from a sampler or instrument in any particular instance is a valid record of the concentration over the period of time at the point of sampling.

It is recognized that the distribution of concentrations of substances in atmosphere at a place of work can vary widely. A statistical approach to such variation is outside the scope of this chapter and is described elsewhere.[45] In some countries where there is a limit for the general concentration of a substance in a workplace atmosphere it is a requirement to apply statistical analysis to the measurements made, with a view to assessing the probability of the general limit being exceeded.[46]

However, the emphasis under COSHH is upon personal exposure. The purpose of measurement is to ensure that the average concentration to which an employee is exposed will be below the limit for the period of exposure. It is relatively unimportant that the same employee, when doing the same job on a different day, may be exposed to a different average concentration.

The 'Approved Code of Practice', paragraph 66, says that normally sampling will be in the breathing zone by means of personal sampling equipment. Nevertheless, because of errors in sampling and analysis such as have been mentioned, there may be differences between the amount inhaled and the concentration measured. In addition, since the concentration at the workplace is likely to be inhomogeneous, there may well be real differences between concentrations measured and those inhaled. The sampler may be located for longer periods closer to the source of the substance than is the nose or mouth of the individual wearing it: several experiments have shown differences between concentrations measured by identical samplers attached to two lapels of the same operative. A recent systematic study of error in the measurement of total inhalable dust in many dusty places of work showed that on 95% of occasions the measurement taken on one lapel was between half and twice that taken on the other. The variation in atmosphere of gases and vapours is likely to be less than that of dust but this study suggests that, in order to be more than 97.5% certain that a sample for total inhalable dust taken from one lapel will exceed the exposure limit, the sample taken at the same time from the other lapel must show more than twice that limit.

These conclusions indicate the importance, when concentrations are in the region of or above an exposure limit, of taking more than one measurement. To refer again to the example above, if two measurements were taken then both must be over the limit by 1.4 times to give roughly a 97.5% certainty that it was exceeded: for three measurements, the corresponding factor would be 1.15.

It will be seen from the considerations outlined in this chapter that the measurement of substances in atmosphere is not a simple matter which can be undertaken by an untutored or inexperienced person. The selection of equipment, instrumentation, and laboratory procedures have to be considered carefully, and interpretation of results in terms of the Regulations requires skill and knowledge.

References

1. 'Health and Safety. The Control of Substances Hazardous to Health Regulations 1988: Approved Code of Practice, Control of Substances Hazardous to Health and Approved Code of Practice, Carcinogenic Substances', SI No. 1657, HMSO, London, 1988 (ISBN: 0 11 885468 2).
2. 'Occupational Exposure Limits' Guidance Note EH 40/90, HMSO, London 1990 (ISBN: 0 11 8854208).
3. 'Health and Safety. The Control of Lead at Work Regulations', SI No. 1248, HMSO, London, 1980.
4. 'Threshold Limit Values and Biological Exposure Indices 1989–1990', American Conference of Government Industrial Hygienists, Cincinnati, Ohio, 1989 (ISBN: 0 936712 81 3).
5. 'Maximum Concentrations at the Workplace and Biological Tolerance Values for Working Materials', VCH Publishers, Cambridge, 1989 (ISBN: 0 89573 660 8).
6. Reference 1, paragraph 66.
7. Reference 2, Appendix 1.
8. 'Monitoring strategies for toxic substances', Guidance Note EH 42, HMSO, London, 1989 (ISBN: 0 11 885412 7).
9. 'Methods for Biological Monitoring', ed. T. J. Kneip and J. V. Crable, American Public Health Association, Washington D.C., 1988 (ISBN: 0 87553 148 2).
10. 'Human inhalation pharmacokinetics of 1,2,3-trichloro-1,2,2-trifluoroethane (Fluorocarbon 113)', B. H. Woollen *et al.*, *Int. Arch. Occup. Environ. Health*, 1990, **62**, 73.
11. 'Detection and Measurement of Hazardous Gases', ed C. F. Cullis and J. G. Firth, Heinemann, London, 1981.
12. 'Gas Detector Tubes', British Standard 5343, British Standards Institution, Milton Keynes, 1976.
13. 'Second International Meeting on Chemical Sensors', ed. J.-L. Aucouturier *et al.*, University of Bordeaux, France, 1986 (ISBN: 2 906257 00 1).
14. Health and Safety Executive, Occupational Medicine and Hygiene Labortory, Red Hill, Broad Lane, Sheffield.
15. 'Controlling Airborne Contaminants in the Workplace', Technical Guide No. 7, British Occupational Hygiene Society, Northwood, Science Reviews, 1988 (ISBN: 0 905927 42 7).
16. K. Y. K. Chung and N. P. Vaughan, Comparative laboratory trials of two portable direct-reading dust monitors. *Ann. Occup. Hyg.*, 1989, **33**, 591.
17. P. J. Lioy and M. J. Y. Lioy, 'Air Sampling Instruments', American Conference of Government Industrial Hygienists, Cincinnati, Ohio, 1989 (ISBN: 0 936712 43 0).

18. 'The Selection and Use of Personal Sampling Pumps', Technical Guide No. 5, British Occupational Hygiene Society, Norwood, Science Reviews, 1985 (ISBN: 0 905927 86 9).
19. 'Methods of Air Sampling and Analysis', American Public Health Association Intersociety Committee, Washington D.C., 1972.
20. 'NIOSH Manual of Analytical Methods', U.S. Department of Health and Human Services, Washington D.C., 1984.
21. R. H. Brown and C. J. Purnell *J. Chromatogr.* 1979, **178**, 79.
22. 'Diffusive Sampling—an Alternative Approach to Workplace Air Monitoring', ed. A. Berlin, R. H. Brown and K. J. Saunders, The Royal Society of Chemistry, London, 1987 (ISBN: 0 85186 343 1).
23. R. H. Brown, J. Charlton, and K. Saunders, 'The Development of an Improved Diffusive Sampler', *Am. Ind. Hyg. Ass. J.*, 1981, **42**, 865.
24. A. Bailey and P. A. Hollingdale-Smith, 'A Personal Diffusion Sampler for evaluating time-weighted exposure to organic gases and vapours', *Ann. Occup. Hyg.*, 1977, **20**, 345.
25. E. V. Kring *et al.*, 'A new passive colorimetric air-monitoring badge system for ammonia, sulphur dioxide and nitrogen oxide', *Am. Ind. Hyg. Ass. J.*, 1981, **42**, 373.
26. K. Pannwitz, Direct reading diffusion tubes, *Drager Review* 1984, **53**, 10.
27. 'Methods for Determination of Hazardous Substances 58, Mercury vapour', HMSO, London (ISBN: 0 7176 0292 3).
28. Proceedings of the 1959 Pneumoconiosis Conference, Johannesburg, ed. A. J. Orensten, Churchill, London, 1960.
29. 'Methods for Determination of Hazardous Substances 14, Total inhalable and respirable dust gravimetric', HMSO, London (ISBN: 0 7176 0142 0).
30. D. Mark, 'Problems associated with the use of membrane filters for dust sampling when compositional analysis is required', *Ann. Occup. Hyg.*, 1974, **17**, 35.
31. 'Methods for Determination of Hazardous Substances 39/2, Asbestos fibres in air', HMSO, London (ISBN: 0 7176 0300 8).
32. 'Handbook of Occupational Hygiene', ed. B. Harvey *et al.*, Croner Publications, New Malden, 1990 (ISBN: 0 903393 50 6).
33. 'Identification and Analysis of Organic Pollutants in Air', ed. L. Kieth, Butterworths, London, 1984 (ISBN: 0 250 40575 X).
34. 'Methods for Determination of Hazardous Substances'. See: Bibliography.
35. N. P. Vaughan *et al.*, 'Filter weighing reproducibility and the Gravimetric Detection Limit', *Ann. Occup. Hyg.*, 1989, **33**, 3313.
36. 'EH 40 Workplace Air and Biological Monitoring Database', HMSO, London, 1990.
37. 'Measuring up to the Competition', Cmd. 728, HMSO, London, 1989.
38. 'Methods for Determination of Hazardous Substances 27, Diffusive sampler evaluation protocol', HMSO, London, 1987 (ISBN: 0 7176 0287 7).
39. 'Methods for Determination of Hazardous Substances 54, Protocol for assessing the performance of a pumped sampler for gases and vapours', HMSO, London, 1986 (ISBN: 0 7176 02753).
40. 'Documentation of NIOSH Validation Tests', U.S. Department of Health and Human Services, Washington D.C., 1977.
41. 'Council Directive of 16 December 1988 amending Directive 80/1107/EEC', *Official Journal of the European Communities*, No. L356/74 to L356/78, Brussels, 24 December 1988.

42. N. P. Crawford and A. J. Cowie, 'Quality Control of Airborne Asbestos Fibre Counts in the U.K.—the Present Position'. *Ann. Occup. Hyg.*, 1984, **28**, 391.
43. 'Methods for Determination of Hazardous Substances 70, Analytical quality in workplace air monitoring', in press.
44. 'Workplace Analysis Scheme for Proficiency—Information for Participants', Health and Safety Executive Committee on Analytical Requirements, Occupational Medicine & Hygiene Laboratory, 403/405 Edgware Road, London, NW2 6LN, 1988.
45. R. P. Harvey, 'Statistical Aspects and Air Sampling Strategies' (In Reference 11, *q.v.*).
46. 'NIOSH Handbook of Statistical Tests for Evaluating Employee Exposure to Air Contaminants', HEW Publication No. 75-147, U.S. Department of Health and Human Services, Washington D.C., 1975.

CHAPTER 5

Statutory Records

P. J. Hewitt

Introduction

Employers have a duty of care under Common Law and also statutory obligations (for example, under Section 2 of the Health and Safety at Work Act) to provide safe systems of work. Each employer must be able to demonstrate such a commitment by means of company records and other documents. The COSHH Regulations specify five basic statutory records, as detailed in Table IV.

Before going on to discuss these more fully it is important to appreciate that statutory records form only part of an integrated system for health and safety. Most establishments prepare documentation for a variety of purposes and it is both economical and efficient whenever possible to refer to existing records. As an example of this, most establishments have a 'Goods Inward' section in which records are maintained of the identity and quantity of substances delivered to the site, and of the destinations of these substances when there. Such records may well provide a basis for an inventory of substances.

It is likely that there will be written procedures for standard operations in production departments and elsewhere, and these can be invaluable as references for the relevant COSHH assessments. With the motivation of requirements for quality management as set out in British Standard 5750, many manufacturing and service organizations have documentation and record systems under regular review—thus affording opportunities to integrate with them such records as are required for purposes of health and safety.

TABLE IV *Basic statutory records required by the Regulations*

Reference in the Regulations	*Type of record*	*Details (including those in Approved Code of Practice)*	*Period of time for retention* (years)
6	Assessments, in all but the simplest cases	Kept by employer; to be reviewed as necessary	
9(4)	Examinations and tests of engineering controls and of respiratory protective equipment	Kept by employer; to include defects, repairs, and other information as specified in ACOP	five
10(3)(a)	Reports of monitoring at place of work under Regulation 10(1) and of personal exposures of identifiable employees	Kept by employer: to include certain information specified in ACOP, to be retrievable easily, and comparable with requirements under Regulation 11	thirty
10(3)(b)	Reports of monitoring at place of work under Regulation 10(1) in any other case	As above; as appropriate	five
11(3)	The health of each employee who may be required under Regulation 11(1) to be under surveillance	Kept by employer; to include information as approved by the Health and Safety Executive	thirty

Note

Discretion may be exercised in selecting a format for records; they must be easy to retrieve, with information in a form that is clear and easy to understand.

Review of Existing Records and Arrangements

(i) Inventory of Substances

Individuals and organizations that supply substances of various kinds have duties under Section 6 of the Health and Safety at Work Act to make available technical information about the safe use of their products: amendments under the Consumer Protection Act 1987 require that relevant information of this nature be supplied to the purchasers.

Every establishment should maintain records of incoming goods and the objective should be to ensure that these are comprehensive and that every substance entering a site is logged. If micro-organisms are involved, record systems should be modified to allow for this. Hazard data sheets should be available for each of the materials concerned and should be distributed within the organization so that the operators concerned have access to all necessary information.

The collation of these data sheets is an important aspect of the procedure for assessment. When chemical products are transferred elsewhere—off the site for further processing or to the ultimate customer—they should be accompanied by the correct safety documentation.

Records of the disposal of waste do not form a part of this chapter but they are of necessity part of the integrated system of documents that is desirable. The disposal of all wastes—and particularly 'Special Waste' as defined in relevant legislation—is controlled by the authorities and every establishment's integrated record system should include balancing data covering materials consigned in this form.

(ii) Manufacturing and Laboratory Procedures

Mention has been made already of existing written procedures. Usually their main purpose is to ensure that standard practices are followed irrespective of changes in personnel. Although not specifically concerned with safety they are likely to emphasize the practical importance of keeping to the method given—and in this respect may be particularly helpful to assessment.

Production procedures may refer to the personal protective equipment that should be worn, or to the exhaust ventilation system—in other words, they may at times make specific reference to matters of health and safety.

(iii) Control Systems

Within any organization implementation of the Regulations normally would require a combination of administrative controls and of process and engineering controls. Site audits should be carried out regularly—to check not only that the systems and arrangements for control are sensible and complete but also that in practice they are followed.

Particular problems can arise with regard to local exhaust ventilation and before proceeding with the formalities of assessment in a particular area it can be extremely useful to have detailed information on these systems—including their original purpose, the performance for which they were designed (volumes and rates of flow of air, *etc.*), and their current performance. As described later, records of local exhaust ventilation are required but during the assessment procedure a loss of time and considerable frustration can be avoided if a site-wide audit of such systems has been made and the results and information from it are available from the start.

(iv) Records of Personnel and Health

Aside from the statutory requirements, personnel records can play an important part in the system of documentation. They can assist historically by identifying the areas of an establishment in which individuals have worked, and the dates and periods of such engagements. Personnel or medical departments should have records of any complaints, details of absence because of illness, or other relevant information—such as mention of any history of asthma, of specific allergies, or of any known laxity, indiscipline, carelessness, or objection to the use of personal protective equipment.

While on the subject of personnel records in particular it is important to ensure so far as practicable that all such data are kept in standard form and that there is overall consistency in the approach (in other words, that the same information is always given the same weight). Procedures for compiling and maintaining records should be set out in detail and those responsible for carrying out the work should be trained, fully familiar with them and required to follow them strictly.

It should be remembered also that any system which records personal data on a computer must be licensed by the Registrar for Data Protection.

(v) Communication

Maintaining a good two-way flow of information is an important aspect of successful management in any organization, large or small: however, in practice it is found that audits of systems of communication frequently indicate inadequacies.

Accurate and easily understood records are integral parts of management for health and safety, as also is effective communication.

Table V on page 71 gives examples of notifications and other transfers of information which are required under the Regulations. Records are to be made available on request by the Health and Safety Executive and by certain other authorities; in specified cases duties are placed on the employer to communicate the content of records to employees and to their safety representatives.

Implications Under Civil and Criminal Law

It is suggested that the prudent employer will do more than keep the statutory records required under the Regulations, and that these should form an important part of an integrated system of documentation. At a hearing for an alleged breach of statutory duties, in addition to those papers specifically required, a court might well take into account the submission of documentary evidence with regard to (say) the systems of work. Such evidence could be a powerful aid to a defence under Regulation 16, as an indication that the employer did take all reasonable precautions and exercised all due diligence to avoid the commission of an offence.

However, it should be noted that such a defence under Regulation 16 is available only in criminal proceedings. It is anticipated too that COSHH will figure prominently in future in civil claims in the courts. In such claims in the past it was common to find references to alleged breaches of statutory duties under the Factories Acts (such as of Section 63 on ventilation, now repealed under COSHH) but now it may not be an adequate defence if an employer were to argue that checks of a ventilation system were made each year, that appropriate records of these were kept, and therefore due diligence was

TABLE V *Notifications and other transfers of information required under the Regulations*

Reference in the Regulations	*Action required (including as stated in Approved Code of Practice)*	*Time limit given*
12(1)	The employer to inform employees or their representatives of the results of assessments	
11(4)	If his activities come to an end, an employer holding health records must notify the Health and Safety Executive	forthwith
11(6)	An entry in a health record by an employment medical adviser or the appointed doctor that an employee should not be engaged on work specified	
11(7)	An entry in a health record by an employment medical adviser or the appointed doctor that surveillance of an employee must continue after he or she ceases to be engaged in work specified	
11(8)	The employer to provide an employee with his personal health record	reasonable notice
11(9)	An employee to provide information about his or her health to the employment medical adviser or appointed doctor	
11(11)	In accordance with a procedure approved by the Health and Safety Executive, application to the Executive when an employer or an employee is aggrieved by a decision given in a health record	within 28 days of notification of the decision recorded
	Notification to the employer or employee of the result of a review of an entry in the health record	in accordance with approved procedure
12(1)	Employer to provide adequate information, instruction, and training for employees as to the nature of substances used, risks to health and precautions to be taken	
12(2)(a)	Employer to provide information on the results of monitoring required under Regulation 10, and in particular to inform employees or their representatives if the results show that a maximum exposure limit was exceeded	
12(2)(b)	Employer to inform employees of collective anonymous results of any health surveillance required under Regulation 11	
12(3)	Employer to ensure that any person (whether or not his employee) who carries out work in connection with the employer's duties has the necessary information, instruction and training	
13(2)(a)	Employer to notify in advance certain fumigations intended, as specified in Schedule 7 Part I (Persons to be notified) and Part II (Information to be given)	twenty-four hours before, or shorter periods by agreement
13(3)	Employer to ensure that suitable warning notices have been fixed at all points of reasonable access to the premises before fumigation, and that they are removed afterwards	beforehand and afterwards
14(1)	If appropriate, the Health and Safety Executive to issue exemption certificates	on request

exercised. The Regulations are wide-ranging and they open new and so far untested possibilities.

An important part of the claims procedure is the discovery of documents, which requires an exchange of all relevant documents between the parties to a dispute. It will be appreciated that this will include not only statutory records but also—as examples—written procedures, information about hazards, audits, and other internal reports; the possibility that at some time documents might be required for this purpose underlines the desirability of having an extended and integrated system that will be both reliable and consistent.

The selection of a method of keeping records and the design of the system are matters for the employer and will depend to a large degree on the size and function of his unit. However, whatever approach is used it is important that the format should present all the information clearly in a manner that can be understood. If a computer is employed the records should be easy to retrieve, and up-to-date printed copies always should be available.

Procedure for Assessment

Normally, as a preliminary to making an assessment, it is necessary to make one or more preparatory visits to the place of work—with a view, for example, to identifying the areas in which detailed formal examination will be necessary. At the first visit it should be possible also to sketch (or, with permission, to photograph) the basic plan of the office or factory, to ascertain what records exist (as examples, of maintenance of plant, or data sheets on the substances used), and whether there are deficiencies. In some cases already there may be records of past investigations into air quality or of other working conditions, made by inspectors or by scientists commissioned for the purpose.

After preparation on lines such as these, a second or subsequent visits can be made in the light of established fact and with particular reference to any aspects unresolved hitherto. Again, the detail of the work will differ as a reflection of the type of unit and the activities carried on there, but the check list given below can be used as a guide and modified as necessary.

(*a*) *Check that normal working is in progress*—record the number of operators engaged, numbers of support staff and supervisors; note whether output is at maximum (that all lines and machines are working at maximum speeds); note any malfunctioning of equipment or of plant, or other unusual conditions; through all stages of the check list, note the presence of any personnel from contractors or maintenance departments.

(*b*) *Evidence of contamination*—note whether dust is visible in the air, on surfaces, on face, hands, or clothing of employees; note any fume or vapour visible, leakages or spills of material.

(*c*) *Personal facilities*—see changing areas and note practice in their use; examine similarly facilities for washing and eating; check access to medical services.

(*d*) *Work practices*—note practices in use of substances and precautions to ensure safe handling; check knowledge of precautions required with operatives and their supervisors; check whether there is direct physical contact with a substance; check training in procedures and the proper use of control equipment.

(*e*) *Ventilation systems*—check type of system and arrangements for ventilation, including location, condition, and standard of cleanliness of air inlets and outlets; if local exhaust ventilation is employed, check similarly; note in all cases air flows and general efficiency.

(*f*) *Personal protective equipment*—note type and manufacturer concerned; check procedure for supply to operatives and whether it is appropriate for the purpose and used as it should be; note any complaints, including (especially in the case of respiratory protective equipment) any evidence of poor fit; check maintenance and training in procedures for proper use.

(*g*) *Storage facilities*—note the method of storage of the substance concerned (including whether containers are open or sealed, and how they are protected against damage); check the possibility of exposure of personnel in a storage area, either normally or as a result of accident; check adequacy of ventilation; any special procedures (such as for spills, fire, or explosion); check methods of transfer within the store; control procedures for drawing substances from store; check methods of transfer, including type and condition of the containers used.

(*h*) *Working procedures*—note how each substance is used; check the possibility of escape during dispensing, manipulation, and all other stages; note any signs of leaks or spills; check the method of retaining or transfer of each substance when it is not in use.

(*i*) *Cleaning and disposal of waste*—note cleaning methods employed (including whether wetting, solvents, or vacuum are involved); any special procedures; whether the cleaning of floors, surfaces, clothing, *etc*., uses air blown under pressure; check procedures for collection and disposal of waste.

(*j*) *Emergency facilities*—note procedures for dealing with leakage and spills, including the availability of appropriate equipment; note availability and condition of emergency equipment; check plans for emergencies, including whether and how operatives are informed about these plans; note any possibility for the exposure of people from elsewhere (that is, not employed in the unit).

In most circumstances, by following stages such as these in his task of gathering data and inspecting the place of work an assessor should obtain enough information to complete the necessary report or reports. In more complex situations more information may be needed, for example, by making use of air sampling.

It is stated in the Approved Code of Practice that in the most simple and obvious cases, which can be repeated easily and explained at any time, a written or recorded assessment is not needed. However, this is qualified in various ways and it would be prudent to keep written reports of all assessments made, together with related documents.

A report of an assessment should be concise. Supplementary information can be appended and (with a view to avoiding repetition) cross-references made to other documents. The reports will vary in detail according to circumstances but in general should include:

(*a*) *The purpose and scope*—indicating whether the document is intended to be a comprehensive new assessment, part of an assessment, or a re-assessment

(*b*) *Place, date, and time*—including also precise identification of the substance, process, or unit that is being investigated

(*c*) *Inventory of substances*—a list of materials present at the location, with note of the substance or substances being assessed

(*d*) *Summary of process*—a description of normal operations with note of any changes observed or anticipated which could affect accuracy of assessment

(*e*) *Personnel*—the names or other identification of managerial staff, supervisors, operators, and ancillary workers concerned

(*f*) *Evaluation*—possible routes of exposure to the substance or substances; procedure for monitoring and conclusions drawn (with data appended); measures for control (with information appended on the performance of local exhaust ventilation); the extent of supervision, training, and skills of operators; other administrative aspects; the standards applied (exposure limits, Health and Safety Executive 'Guidance Notes', other national or international specifications—with full details in appendices)

(*g*) *Conclusions*—an assessment of the risks to health, and the steps which need to be taken to comply with the COSHH Regulations

(*h*) *Recommendations*—including the circumstances in which re-assessment would be required

(*i*) *Documents cited*—(as references in the report, with details appended)

(*j*) Signature, date, position, and qualifications of the assessor

(*k*) Signature, date, and position of the employer accepting the assessment.

The employer has a responsibility to appoint a competent assessor but responsibility for the actual assessment remains with the employer and he should indicate in writing on each one that he accepts it. A large organization might establish a procedure for acceptance of assessments—for example, through a director of the company with special responsibility for health and safety who can speak for the organization as a whole.

The employer may consider the use of a standard format for records of assessment and for other documents under COSHH, and many such products are offered for sale. These goods (briefly stated, either special stationery or computer programmes) can be useful—particularly for simple cases—but especially for the records of assessment it could well be advisable to develop a format for the organization, to be compatible with other systems in use.

Examinations and Tests of Control Measures

Requirements for maintenance, examination, and tests of measures to prevent or control exposure are set out in Regulation 9.

Thorough and regular examination and tests of engineering controls are specified. Maintenance includes: visual checks; inspections; testing; preventive servicing; and remedial work. Similarly, at suitable intervals, thorough examination and (where appropriate) tests of respiratory protective equipment are required.

Many different people may be involved in work of this nature—not only maintenance staff.

(i) Engineering Controls

This section places emphasis on local exhaust ventilation but all forms of engineering control require regular examination. (For example, a check on the efficiency of a thermostat can prevent accidental release of excess solvent from a dryer.) Many routine examinations can be carried out easily and simply, with ready observation of obvious faults such as blocked air exhausts or missing sections of partial enclosures, but comprehensive examinations and tests such as of local exhaust ventilation units require specially trained people.

In all maintenance it is essential to set up systematic procedures and to keep good records. Schedule 3 of the Regulations gives minimum frequencies and tests for four processes and otherwise LEV plant should be examined and tested at least every fourteen months. It may be necessary to test older plant rather more frequently than new. Quite detailed measurements must be made to provide data on which the continuing efficiency (or otherwise) of a unit can be judged; information required could include:

(*a*) *Enclosures or hoods*—a statement of the maximum number of these to be in use at one time; their locations; the static pressure behind each hood or at point of extraction; velocity over face

(*b*) *Ducting*—dimensions; transport velocity; volume and rate of flow

(*c*) *Filter or collector*—specification; static pressures at inlet, outlet, and across filter; volume flow

(*d*) *Fan or air mover*—specification; static pressure at inlet; volume flow; direction of rotation

(*e*) *Systems that return exhaust air to the place of work*—efficiency of filter; concentration of contaminant in air returned.

A record for each thorough examination and test of an LEV unit should show:

(*a*) Name and address of employer responsible for the unit

(*b*) Identification and location of the unit, together with the process and substance or substances concerned

(*c*) Date of the previous thorough examination and test

(*d*) Conditions at the time of the test (as examples, whether under normal production or special conditions—whether being used at maximum or out of commission

(*e*) Information about the LEV unit showing: the operating performance required for control of the substance under Regulation 7; whether this performance is achieved; if not, the repairs required in order to obtain it; the methods used in order to make these judgments (visual, measurements of pressure or air flow, dust lamp, air sampling, tests of filter integrity, *etc.*)

(*f*) Date of examination and test

(*g*) Name, designation, and employer of person carrying out the examination and test

(*h*) Signature of person carrying out examination and test

(*i*) Details of repairs carried out (to be completed by the employer responsible for the plant); the effectiveness of the repairs should be proved by a re-test.

(ii) Personal Protective Equipment

Personal protective equipment includes: eye protection; footwear (boots, safety shoes, *etc.*); gloves; protective clothing (aprons, overalls, special suits, *etc.*); and respiratory protective equipment ('RPE'). Basic records should be kept for each type of equipment, including dates of issue to employees and maintenance. The Approved Code of Practice gives requirements for records for RPE (other than one-shift disposable masks), which should cover:

(*a*) Name and address of employer responsible for the RPE

(*b*) Particulars of the equipment and reference number

(*c*) Report on the condition of the equipment (the integrity of straps, face-pieces, valves, filters, *etc.*), and particulars of defects

(*d*) Report on any air supply system to the RPE, including checking pressure in supply cylinders and, in the case of apparatus fed by airline, the volume, flow, and quality of the air supplied (such as whether it was free from oil and other contaminants)

(*e*) Date, name, and signature of person carrying out the examination and tests.

Monitoring of Exposure at the Place of Work

Regulation 10 requires that where necessary the exposure of persons to substances hazardous to health be monitored and the results recorded. Normally, monitoring necessitates a series of tests of the environment at the place of work for substances concerned, with consideration of the results in relation to possibilities such as inhalation, absorption through the skin, or the ingestion of harmful material. The objective is to obtain information about such exposures expressed in quantitative terms.

Commonly, air sampling will be carried out in the immediate environment or breathing zone of the operator, with a view to ascertaining whether there is risk by inhalation. In other cases, air sampling may be to check the effectiveness of control measures. Further approaches which can be useful at times include the analysis of settled dust, or of other contamination on surfaces, and checking of risk of ingestion from transfer on the hands or via clothing. 'Biological monitoring' (the analysis of body fluids, or other specimens) may be appropriate for certain types of work.

In all cases appropriate records of monitoring of exposure must be kept. In the case of personal exposures the actual record or a suitable summary must be retained for thirty years; in other instances, for five years. Normally the record will be the report or a compilation from reports of the laboratory carrying out the work; such a laboratory may be a part of the organization concerned or external. The extent of the information contained in the report (or summary) will depend on a variety of factors but certain minimum requirements in relation to personal air monitoring are listed below:

(*a*) Identification of the premises (address) and the precise site (location at the unit) that was examined
(*b*) Note of the process, including whether it was in normal operation, under maintenance, or any other unusual circumstances (loss of power, breakages, *etc.*) during sampling
(*c*) Name or names of those monitored, with notes of the work they did
(*d*) Substance or substances monitored for; procedures used
(*e*) Dates, times of commencement, and the periods of time over which samples were taken
(*f*) Results, with units in which they were expressed (*e.g.* mg m^{-3}, with limits of accuracy), and interpretation (*e.g.*, in comparison with eight hour time-weighted average reference values for the substances concerned)
(*g*) Signature of person responsible for the results, reference number of report, name and address of laboratory, date.

In most cases the frequency of monitoring should be at intervals of not less than twelve months. For vapour or spray from certain chromium electrolytic processes specified in the Regulations monitoring must be at least every fourteen days. For certain types of work involving vinyl chloride monomer it either must be continuous or in accordance with a procedure approved by the Health and Safety Commission.

Health Surveillance

A record with the following particulars should be kept for every employee undergoing health surveillance:

(*a*) Surname, forenames, sex, date of birth, permanent address, post code
(*b*) National Insurance number
(*c*) Date of commencement of present employment
(*d*) Historical record of work in this employment involving exposure to substances requiring health surveillance

(*e*) Conclusions of all other health surveillance procedures, by whom carried out, with dates.

In such records the conclusions should not include confidential clinical data: they should give a view as to the fitness for work of the employee concerned by an occupational health nurse or other person with suitable qualifications, and where appropriate the decision of the appointed medical doctor or employment medical adviser.

If the requirement under Regulation 11(3) is to keep only a record of health, item (*e*) above need not be included. An employer must keep any such information for thirty years; if his activities come to an end he must notify the Health and Safety Executive in writing of the fact, and offer to it the records concerned.

In addition to the foregoing, the employer should keep an index or list of persons under health surveillance in a form that can be related to the records of monitoring required by Regulation 10. The object here is so that both soon after and long after an exposure might have taken place its nature and extent can be related to any symptoms noted within the group of employees.

Normally, the full clinical records would be maintained by the medical practitioners concerned and would be subject to the customary confidential relationship with a patient. Extracts or conclusions drawn from them would be included as necessary in documents required under the Regulations.

Training, Notifications, and Other Transfers of Information

Section 2(2)(c) of the Health and Safety at Work Act 1974 places upon an employer a general duty to provide 'such information, instruction, training and supervision . . . to ensure . . . the health and safety at work of his employees'. The Regulations reinforce this duty and invoke also many specific notifications and other transfers of information. Every employer should keep suitable records to show that he has recognized and acted upon these requirements.

Regulation 12(1) requires an employer to provide information, instruction, and training so that employees will know of a risk to health and of precautions to be taken. The details set down should include, as a minimum:

(*a*) Name or names of the employees who took part

(*b*) Date of provision of the information, instruction, and training

(*c*) Summary of the topic and field or fields covered.

Supplementary records will be needed for specific training in matters such as the use of respiratory protective equipment. In all cases, the employees participating should sign to show that they did so, with careful filing of the relevant reports in case these may be required at any time in the future.

Table V shows other transfers of information necessary under the Regulations and there are requirements for notifications (such as warning in advance of certain procedures in fumigation). Employees or their representatives must be informed of the results of assessments and of any monitoring required under Regulation 10, particularly if a maximum limit was exceeded. Collective results (in a form that preserves the anonymity of individuals) must be provided for any health surveillance imposed under Regulation 11.

Acknowledgement. This paper included tabulations and summaries which are based on parts of 'The BURL Guide to the Control of Substances Hazardous to Health Regulations 1988'; that original material is copyright by the author, Peter J. Hewitt, and its use here was with the permission of H & H Scientific Consultants Limited, Leeds.

CHAPTER 6

Recording COSHH Assessments

P. HOPKIN

Introduction

The chapter will discuss various practical implications of the need to record assessments, such as:

1. Circumstances in which the assessments have to be produced as written records
2. Nature and extent of information which should be recorded
3. Options available to employers
4. Various formats which are available and their application in different circumstances
5. Use of the written assessments so that continuing requirements for record keeping may be identified
6. How the written assessments will assist employers in complying with the requirements of Regulations 7 to 12.

Written Assessments

Regulation 6 lays down the requirement to undertake assessments. The Regulation is set out as follows:

> An employer shall not carry on any work which is liable to expose any employees to any substance hazardous to health unless he has made a suitable and sufficient assessment of the risks created by that work to the health of those employees and of the steps that need to be taken to meet the requirements of these Regulations.

The Approved Code of Practice (ACOP) under the Regulations[1] offers guidance on the meaning of 'suitable and sufficient' in the context of Regulation 6. The Health and Safety Executive and several other organizations, including trade associations, have published detailed advice and guidance on the factors which should be taken into account when undertaking assessments of specific processes.

A later part of this chapter will consider the extent of information which should be recorded so that it will be a 'sufficient' assessment under the Regulations.

In most cases, it will need to be recorded in writing so as to be considered 'suitable'.

The ACOP states that 'in the simplest and most obvious cases which can be repeated easily and explained at any time an Assessment need not be recorded'. An assessment may be simple and straightforward because the process is low risk, involves limited exposure duration, or may be explained easily and repeated at any time.

Although the Health and Safety Executive has taken the view that certain assessments need not be recorded, much of the information and advice published by Environmental Health departments states clearly that COSHH Assessments always should be produced in writing. Employers with work activities which are subject to health and safety enforcement by the Environmental Health Officers, should consider carefully the advice given by Environmental Health Departments[2] and produce all their COSHH Assessments in writing.

There are as a matter of fact many advantages in doing so. For employers who are concerned with industrial activities, production processes will require written assessments. For industrial employers, the assessments for lower risk activities, such as offices and warehouses, *etc.*, may be recorded with little extra effort.

For employers who are concerned only with low risk activities, the production of written assessments will represent a small additional burden and will prevent adverse comment by the Environmental Health Officers. Also, the production of written assessments will make the task of employee training easier and more consistent. As discussed later in this chapter, the term 'written' refers to the production of a recorded COSHH Assessment. As discussed later, assessments may be kept in document or computer form.

Another advantage of recording *all* assessments is that the reference number allocated to an assessment may be used as a COSHH Code by the Buying Department. Therefore it will be possible to allocate COSHH Codes to all substances purchased and ensure good control of chemicals on the site. When a new substance is to be purchased, the person responsible can decide whether an existing COSHH Assessment will cover the new substance or a new one must be made. In either case, the new substance will be assessed carefully and either absorbed into the existing assessment structure or be allocated a new COSHH Code.

The decision on whether to record COSHH Assessments will depend partly on the advice in the ACOP but also will be influenced by the advice from the local Environmental Health Officers, by the desire of the employer to allocate COSHH Codes to all substances on site, and the advantages associated with having undertaken a full COSHH Audit and being able to demonstrate that all substances have been assessed.

We include in this chapter consideration of information which should be recorded and sample formats for low—and moderate—risk substances and processes. A format is included for a simple office assessment and further formats for assessments of cleaning substances (such as bleach) and also for recording assessments in a laboratory (see pages 88 to 91).

The formats presented have been used successfully by several employers when recording their lower risk assessments. A fuller format will be required for manufacturing hazards. When considering industrial processes it is necessary to assess routine cleaning and maintenance operations in addition to normal production and other normal work activities. The longer format for COSHH Assessments may be based on the Substance Assessment Sheet shown at the end of this chapter. The Process Assessment record will require that some parts of the record form are repeated. The 'COSHH Assessment Grid' and 'Monitoring of Exposure' sections will need to be completed for normal operation and for routine cleaning or maintenance.

Nature and Extent of Information

The Health and Safety Executive has not published a standard format or series of standard formats for recording assessments and therefore employers have the freedom to use their own formats. This flexibility is an advantage for larger employers and for employers who have access to technical information or advisers. However, for smaller employers or those who do not have access to professional safety resources, the absence of standard COSHH Assessment formats may result in confusion and a feeling that the Regulations are vague and difficult to understand.

In deciding on an appropriate format, employers would be well-advised to follow a Risk Management approach towards compliance, undertaking, and recording of assessments. Such an approach is described later in this chapter.

The two main requirements for success in making assessments are:

(i) Accurate identification of all hazards
(ii) Careful description of the element of work which is being assessed.

Obviously an assessment will not be accurate if it fails to identify all the hazardous materials used in and generated by a process. (The correct identification of hazards is described later in this chapter and elsewhere in this book.)

The need to define the element of work being assessed is less obvious. However, it clearly is the case that failure to identify accurately the work activity being assessed will lead to assessments which are too general and which attempt to cover too many activities. As an example, a COSHH Assessment which attempts to describe paint mixing and paint spraying in a single document is not likely to be successful. In general terms, if the precautions necessary during work activities are different, then separate assessments are required or, at the very least, they should be recorded in separate sections of the same assessment.

Preparatory work such as described above is fundamental in determining how many assessments will be needed and the elements of the work to be assessed. The work of assessment is made less onerous if employers can identify in advance hazardous substances and describe clearly the elements of activities being assessed.

Such well-prepared employers may find that a work area which appears superficially to be hazardous and difficult to describe in an assessment turns out to involve quite a small number of activities requiring written assessments. As an instance of this, assessment of an Effluent Plant may appear to be difficult, but in practice it may require only consideration of one or two chemical mixing processes and an assessment of the neutralization treatment itself. In other words, two or three simple assessments relating to the core of the hazardous operation with the remaining work activities classified as 'low risk' and covered by simple safety rules.

The Risk Management approach to undertaking COSHH Assessments has been mentioned. Such an approach requires the following three main steps:

(i) Identification of hazard
(ii) Risk assessment
(iii) Risk response

These steps in turn indicate appropriate formats for recording the assessments.

The elements of Hazard Identification, Risk Assessment, and Risk Response are described later. The sample formats which are shown at the end of this chapter adopt the three steps mentioned. The fuller format for recording process assessments uses separate sheets for recording the steps with a Hazard Identification Sheet for each of the hazardous materials involved in or created by the process. The Substance Assessment Sheet used to record the sample laboratory assessment that is included in this chapter was structured so that the three elements mentioned would be presented on one record form in the correct sequence.

(i) Hazard Identification

Many sources of information are available to help employers with the task of identifying the possibility of hazard associated with a work activity. The main source of such information is the Product Safety Data Sheets which must be supplied with substances purchased.

However, there will be certain circumstances in which this source of information will not be available. For example, Product Safety Data Sheets may not be available for new compounds, and office and cleaning materials rarely come complete with full and detailed safety information (suppliers may have to be pressed for them).

Apart from the possibility of hazard associated with raw materials, employers also will have to take account of hazards which might be generated during a process, or from materials remaining as by-products or unwanted waste. In the case of hazardous materials produced by or during a process, adequate Product Safety Data Sheets may well not be available and employers must look for other sources of advice.

Assistance may be sought from various publications of the Health and Safety Executive, such as the 'Guidance Notes' and the information contained in the 'Authorized and Approved List' published for the 'Classification, Packaging, and Labelling of Dangerous Substances (CPL) Regulations 1984'.[3] The 'Authorized and Approved List' is a particularly useful document because it contains lists of the Risk Phrases and Safety Phrases which may be allocated to various substances.

Guidance Notes have been published on several processes and much useful advice is available from these publications. Guidance Note EH 40 (current issue)[4] lists the substances which have been allocated Occupational Exposure Limits and therefore are subject to the provisions of the Regulations. Often, employers will find that the information available from Product Safety Data Sheets, the 'Authorized and Approved List' and Guidance Note EH 40 need to be summarized and presented in a standard format. This will ensure that relevant data have been obtained and understood. In some cases, it will be necessary to refer to standard texts on toxicology the most commonly used of which probably is 'Dangerous Properties of Industrial Materials', by Sax.[5] Following identification of the hazardous materials which are involved in a work activity, the written COSHH Assessment should list each of these materials and provide a summary of the hazardous properties, including the following information: name of substance; nature of substance; description and container; harmful agent(s); appropriate Risk Phrases; classification under CPL Regulations[3]; and details of Occupational Exposure Limits.

Recording such information provides a basis for the Risk Assessment. The inclusion of Risk Phrases concentrates attention on the particular hazards of each material to be considered in the course of Risk Assessment.

The Substance Assessment Sheet shown at the end of this chapter makes provision on the record form for the information on the hazardous properties of materials.

(ii) Risk Assessment

Identification of the materials and hazards involved are the first steps towards assessment of risk. The Health and Safety Executive has issued a leaflet giving its interpretation of the terms 'Hazard' and 'Risk'. Expressed simply, the *hazard* presented by a substance is its ability to cause injury or ill health. The *risk* presented by the substance takes account of the likelihood of that injury or ill health actually arising in practice. If controlled properly a hazardous substance might give rise only to low risk, while a substance of low hazard (such as a 'nuisance' dust) might be a considerable risk if control measures are inadequate and it is present in substantial quantity.

The purpose of the Risk Assessment, therefore, is to assess the process or work activity and the possibility of harm. Whether an anticipated exposure would be harmful depends on the properties of the material as identified earlier. The Risk Assessment should consider the possibility of harm from the following routes of exposure: skin contact; skin absorption; risk to eyes; inhalation risk; risk of ingestion; and other harmful exposure—such as aspiration or injection.

For each of the routes indicated, a decision should be taken on whether the likelihood of harmful exposure is 'high', 'moderate', or 'low'. This information may be recorded on a 'COSHH Assessment Grid' as shown on the record forms at the end of this chapter.

The Risk Assessment stage should take into account the possibility of defect or failure of the control measures. It may be that monitoring of the exposure is necessary. The Health and Safety Executive has published a Guidance Note on monitoring for toxic substances[6] and this publication describes monitoring on four different levels:

1. Basic qualitative assessment using simple equipment such as smoke tubes or dust lamps
2. Sampling of air in the unit in the course of one or more sessions, using calibrated equipment and subsequent analysis of samples taken
3. Sampling and analysis of air in the unit on a regular basis
4. Special sampling strategies.

Frequently, accurate assessment of risk will require Stage 2 or Stage 3 monitoring.

Risk may arise during routine cleaning and maintenance as well as during normal operation of a process or the normal use of a substance under consideration. As an example, the installation of dust collection equipment should result in much improved conditions during normal operation: however, it will be necessary from time to time to empty and clean the extraction unit and procedures will be needed to prevent exposure to the dust which has been collected. The risks presented by this cleaning operation will be different in character from the risks presented during normal use of the process. A COSHH Assessment Grid and notes on monitoring of exposure should be kept also for routine cleaning and maintenance work operations so that the appropriate control measures may be determined for these routine non-production activities.

When recording the Risk Assessment element of a COSHH Assessment, the following information may be kept:

(i) A list of harmful agents involved in process. (This will be a collective list of the harmful agents noted on the various Hazard Identification Sheets.)
(ii) Safety Phrases should be recorded which are representative of the precautions required for the substances listed.
(iii) The exposure should be assessed using the COSHH Assessment Grid and notes should be included on monitoring of exposure.

(iv) The information should be recorded for normal operation, with a description of the work activity which has been assessed.

(v) Similar information should be recorded for routine cleaning and maintenance (if appropriate), again with a note on the work activity which has been evaluated.

(iii) Risk Response

The purpose of undertaking a Risk Assessment is so that appropriate control measures may be identified and suitable records of these kept.

Control measures may be divided conveniently into three categories:

1. *Engineering Controls*, including Local Exhaust Ventilation, the arrangements for general ventilation, enclosure or isolation of processes and the use of vacuum cleaning equipment for cleaning operations.
2. *Personal Protective Equipment*, including respiratory protective equipment, gloves, eye and face protection. Body protection also should be considered, including overalls and aprons. Wellington boots, gauntlets, *etc*., will be appropriate for certain 'wet' processes (such as chrome plating).
3. *Other Controls*, including provision of emergency showers, equipment for washing eyes and additional washing facilities, requirements for use of barrier cream, availability and use of neutralizing agents, antidotes, and so forth, requirements to obtain authorization or Permit to Work before undertaking particular tasks, procedures for dealing with spills or other emergencies.

When recording the COSHH Assessment, the nature and standard of control measures should be specified. The measures identified should relate directly to the exposures shown on the COSHH Assessment Grid. Therefore, by recording hazard information, assessing exposure, and deciding on control measures a logical step-by-step approach is adopted.

The decision as to appropriate control measures having been taken, the record keeping requirements of the Regulations must be considered. This part of the Risk Response record will not contain actual details of checking and maintenance of control measures: the intention is to record the nature and frequency of the checks and inspections which must be carried out.

The record keeping requirements of the Regulations are detailed and extensive. Many employers have found that the size and scale of the assessment exercise was equalled by the requirements for record keeping. The COSHH Assessments shown at the end of this chapter are for low—or moderate—risk activities where the record keeping requirements may be less extensive than with some industrial processes.

Record keeping needs may arise as follows:

(*a*) Thorough examination and tests of Engineering Controls, including Local Exhaust Ventilation

(*b*) Record of monitoring of exposure, including results of air sampling

(*c*) Issue and maintenance of Personal Protective Equipment, including thorough examination and test of respiratory protective equipment
(*d*) Health surveillance data, including individual health records and results of surveillance procedures
(*e*) Records of training
(*f*) Records of re-assessments and a diary system to ensure that COSHH Assessments remain valid.

Alongside each of these requirements, the frequency with which each is to be carried out should be noted, together with allocation of responsibility for ensuring that the particular actions take place.

When deciding the format for recording an assessment, the degree of risk should decide how extensive the written analysis should be and also the nature and extent of record keeping requirements. For 'high risk' activities, it may be necessary to supplement the basic record with detailed reports on atmospheric sampling, the design and installation of extraction equipment, and other Engineering Controls.

Options for Recording COSHH Assessments

Usually there will be a requirement to keep the COSHH Assessment in permanent form. The two main options are:

1. Manual record
2. Computer record.

The advantages of a manual or paperwork system include simplicity—the records should be easy to access and readable without need for special equipment.

On the other hand, a considerable volume of paper may be involved. To some extent pre-printed forms may be used and these offer advantages in standardization and convenience of storage and retrieval.

The information and advice published by various organizations concerned including the Health and Safety Executive, has been almost exclusively as leaflets or in book form. However, printed matrices are available from a number of suppliers and most employers should be able to find examples appropriate for their needs.[7,8]

On the option of computer records, the choice is much wider—although the variability of the different computer programmes is considerable. A survey in March 1990 of the computer programmes available identified as many as twenty different record keeping packages.

For many employers, the option of recording their COSHH Assessments on computer is not attractive. Product Safety Data Sheets are invariably printed leaflets or booklets and the information given in computer form is unlikely to be beneficial for most organizations.

For employers with activities of low or moderate risk probably the right decision will be to keep records in manuscript form. (Product Safety Data Sheets can then be kept with the assessments to which they relate.)

An advantage of computer records is that a diary function may be

COSHH ASSESSMENT No. 1

OFFICE ASSESSMENT SHEET No. 1 **of** 1

Department Administration **Date** January 1990

Area General Office **No. of People** **Male** 2

Work Activity Clerical work **Female** 6

Name of Responsible Person Supervisor

Substance	Data Sheet Yes	Data Sheet No	Hazard Classification	Potentially Harmful Exposure
Typing Correction Fluid	X		HARMFUL	Risk of inhalation of 111 Trichloroethane
Photograph Spray mount		X	HARMFUL, HIGHLY FLAMMABLE	Risk of inhalation of solvent fumes
Photocopier toner		X	NO RECOGNISED HEALTH HAZARD, UNLISTED	Risk of inhalation of dust/powder

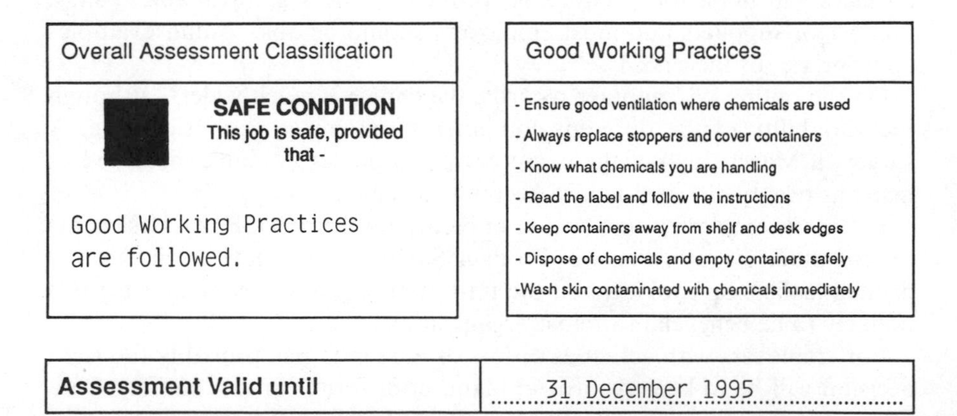

Overall Assessment Classification
SAFE CONDITION This job is safe, provided that -
Good Working Practices are followed.

Good Working Practices
- Ensure good ventilation where chemicals are used
- Always replace stoppers and cover containers
- Know what chemicals you are handling
- Read the label and follow the instructions
- Keep containers away from shelf and desk edges
- Dispose of chemicals and empty containers safely
-Wash skin contaminated with chemicals immediately

Assessment Valid until	31 December 1995

FIGURE 7 *Examples of COSHH Assessment Sheets: (1) Office Assessment*

COSHH ASSESSMENT No. 3.1

CLEANING SUBSTANCE SHEET No. 1 of 5

Department	Kitchen	Date	January 1990	
Area	All areas	No. of People	Male	0
Work Activity	Cleaning activities		Female	2
Name of Responsible Person	Supervisor			
Substance	Bleach	Data Sheet	Yes	☐
Description and container	Liquid in 2 litre plastic bottle		No	X

Harmful Agent(s)	Sodium Hypochlorite
Risk Phrases	R31: Contact with acids liberates toxic gas (chlorine) R34: Causes burns
Safety Phrases	S 2: Keep out of reach of children S28: After contact with skin, wash immediately with plenty of.. water..

Hazard Classification	Potentially Harmful Exposure	Existing Control Measures
CORROSIVE	Risk of skin contact	Gloves, Eyewash bottles

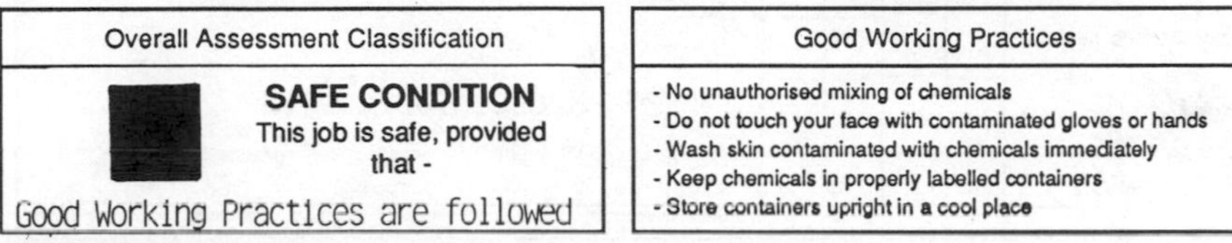

Overall Assessment Classification	Good Working Practices
SAFE CONDITION This job is safe, provided that - Good Working Practices are followed	- No unauthorised mixing of chemicals - Do not touch your face with contaminated gloves or hands - Wash skin contaminated with chemicals immediately - Keep chemicals in properly labelled containers - Store containers upright in a cool place

First Aid Procedures

Skin	Immediately wash or drench affected skin with large quantities of water
Eyes	Immediately wash eyes with plenty of water for at least 15 minutes
Inhalation	Seek medical attention if patient has symptoms apparently due to inhalation
Ingestion	Wash out mouth and give patient plenty of water to drink

Spillage Response

Clean-up Procedure	Absorb in dry sand or absorbent material
Disposal	Wash away to drain with water

Assessment Valid until	31 December 1995

FIGURE 8 *Examples of COSHH Assessment Sheets: (2) Cleaning Substance*

COSHH ASSESSMENT No 4.1

SUBSTANCE ASSESSMENT SHEET No. 1 of 6

Department	Laboratory	No. of people	Male	2
Area	Analytical Sections		Female	1
Work Activity	Use of chemicals			
Name of Responsible Person	Supervisor	Date	January 1990	
Substance	Acids (except hydrofluoric acid)		Data Sheet	Yes ☐ No ☒
Supplier	Laboratory Supplies Ltd			
Description and container	Liquid in 2.5 litre glass Winchester flasks			

Harmful Agent (s)	Dilute mineral acids
Typical Risk Phrases	R34: Causes burns R35: Causes severe burns R37: Irritating to respiratory system

Hazard Symbols and Information	CORROSIVE	OCCUPATIONAL EXPOSURE LIMIT OEL	Ingredient: Sulphuric acid LTEL: (Long Term Limit) 1 mg/m^3 STEL: (Short Term Limit) MEL (Maximium) ☐ OES (Standard) ☒
			Ingredient: Hydrochloric acid LTEL: (Long Term Limit) 5 ppm STEL: (Short Term Limit) 5 ppm MEL (Maximium) ☐ OES (Standard) ☒

Storage Arrangements	Acids cupboard		
Amount in Store	About 10 winchester flasks		
Frequency of use	Daily	Annual Usage	50 ltr
Nature of use/substance	Etching solutions		

Typical Safety Phrases	S 2: Keep out of reach of children S23: Do not breathe gas/fumes/vapour/spray S26: In case of contact with eyes, rinse immediately with plenty of water and seek medical advice S30: Never add water to this product
Assessment Comments	No heating of acids should occur. Sampling tubes failed to detect hydrochloric acid in atmosphere.

FIGURE 9 *Examples of COSHH Assessment Sheets: (3) Laboratory Substance*

COSHH ASSESSMENT No ....4.1.......

Potentially Harmful Exposure
(without control measures)

	High	Mod	Low
Skin - contact		X	
- absorption			X
Eyes		X	
Inhalation			X
Ingestion			X
other exposure			

Existing Control Measures

Overall Assessment Classification

SAFE CONDITION
This job is safe, provided that -

Precautions are taken to avoid splashes

Good Occupational Hygiene Practices are followed

Record Keeping

Subject	Requirement Yes	Requirement No	Frequency	Responsibility	Notes
Engineering Controls (LEV/Extraction)	☐	☒			
RPE/Respirators	☐	☒			
Other Protective Equipment ...As above...........	☒	☐	/	Technician	Dispose of damaged items
Training	☒	☐	Annual	Supervisor	Instructions in Good Laboratory Practices

First Aid Procedures

Skin	Immediately wash or drench affected skin with large quantities of water
Eyes	Immediately wash eyes with plenty of water for at least 15 minutes
Inhalation	Remove patient to fresh air Keep patient warm Seek medical attention
Ingestion	Wash out mouth and give patient plenty of water to drink

Actions in Case of Fire

Extinguishing Agents	Extinguish with dry chemical Extinguish with foam
Special Procedures	In a fire, hazardous decomposition products may be given off

Spillage Response

Clean-up Procedures	**Absorb in dry sand or absorbent material** Remove to safe place
Disposal	Wash small spillages away with water

Assessment Valid until	31 December 1995

introduced, so that re-assessment and other follow-up actions may be prompted by the computer. Also, the extent of information which may be retained is considerable. Further, the information may be cross-referenced with the Buying Department. This helps to ensure that information is readily to hand for all substances purchased and that appropriate action can be taken with regard to new purchases: (a review of the COSHH Assessments can be undertaken and a decision made as to whether the new substance is described accurately by an existing assessment or whether a new one is needed: if an existing assessment is relevant, then the appropriate COSHH Code can be allocated and the substance taken safely into use).

Format for COSHH Record Keeping

When deciding the correct means for recording assessments and, in particular, making the decision between a paperwork system and computer records, employers should bear in mind the objectives which the records should achieve.

The following aspects should be covered:

(*a*) Correct identification of work activity and the substances used
(*b*) Adequate recognition of the hazards presented by the various substances in use
(*c*) Correct statement of the Risk Assessment and adequate identification of exposures
(*d*) Accurate identification of the type and standard of control measures which are required
(*e*) Specification of the records which need to be kept.

The COSHH Assessments shown in Figures 7–9 address the various issues listed above and may form a convenient source of reference and a useful training guide. Other elements that it might be useful to include would be summaries for employees and emergency actions appropriate to the substances.

For summary sheets in particular, several employers have found that it is convenient to use pictograms in compliance with British Standard 5378 'Safety Signs and Colours'.[9]

Again, the format of the 'assessment summary' can be designed, using the British Standard safety signs. The overall assessment may be presented in terms of:

Green Square — Safe Condition
Yellow Triangle — Warning
Red Circle — Prohibition

—and can be reinforced with a presentation of simple safety rules or requirements for a Safe System of Work.

The Assessment Summary Sheet, together with an Emergency Actions Sheet, may be provided for employees on the safety notice board or adjacent to the work area.

Whatever format is adopted, employers must ensure that the COSHH

Assessments and associated records are kept as working reference documents.[10] The Safety Policy also must be a working document and organizations which are successful in keeping their Safety Policy up-to-date and relevant to their work activities undoubtedly will succeed in achieving the same for their COSHH Assessments.

Using COSHH Assessments

To summarize the foregoing, the records of the various assessments should achieve the following objectives:

1. Establish a discipline that will ensure the accurate identification of hazardous materials
2. Maintain good management of hazardous materials so that new substances are classified before they are used at the site
3. The COSHH Assessment recorded should be sufficient to justify the control measures that are in operation
4. There should be enough information in the assessments to confirm that the correct balance exists between engineering controls and the use of Personal Protective Equipment. Where there is undue use of Personal Protective Equipment, an assessment should make this clear and indicate remedial actions
5. The COSHH Assessment should lead to good control of hazardous substances. It is not an end in itself—merely the means by which the steps necessary to achieve good control are identified
6. Suitable assessments will be a powerful training aid
7. The assessments must identify the record-keeping requirements necessary to comply with the Regulations in the future.

References

1. 'Control of Substances Hazardous to Health Approved Code of Practice', HMSO, London (ISBN: 0 11 885468 2).
2. See, for example, 'Sample COSHH Assessment form', Wolverhampton Metropolitan Borough Council, Environmental Health and Consumer Services Department, Civic Centre, St. Peter's Square, Wolverhampton, WV1 1RS.
3. 'Authorized and Approved List. Information Approved for the Classification, Packaging and Labelling of Dangerous Substances for Supply and Conveyance by Road', HMSO, London (ISBN: 0 11 883901 2).
4. 'Guidance Note EH 40/91 Occupational Exposure Limits 1991', HMSO, London (ISBN: 0 11 885580 8).
5. N. I. Sax and R. J. Lewis Senior, 'Dangerous Properties of Industrial Materials', Van Nostrand Reinhold, New York, 1988 (ISBN: 0 442 31813 8).
6. 'Guidance Note EH 42 Monitoring strategies for toxic substances', HMSO, London (ISBN: 0 11 885412 7).
7. 'Record Keeping Book for COSHH', Croner Publications, New Malden.
8. 'COSHH Assessment Workbook', Institution of Occupational Safety and Health.

9. 'British Standard 5378 Part 1: 1980. British Standard 5378 Part 3: 1983', British Standards Institution, Milton Keynes.
10. See, for example, 'COSHH Regulations 1988', Register, Progressive Risk Assessments Limited, 17 Church Road, Northfield, Birmingham, B31 2JZ (ISBN: 0 9515698 0 5).

CHAPTER 7

Microbiological Aspects

C. H. COLLINS, MBE

Introduction

The implementation of the COSHH Regulations initiated a new industry—the publication of many books and papers and the provision of courses of instruction on how to apply them. In all of this excitement the microbiologist appears to have been overlooked. Useful guidance came from the Health and Safety Executive and the chemical industry about carcinogens, hazardous chemicals, and other materials, but there has been little, if anything, that might help the microbiologist and those who use micro-organisms in laboratories and in industry. While it is not difficult to identify and to quantify poisonous and hazardous chemicals, to measure the amounts of vapours, gases, and dusts in the atmosphere of a workplace (the techniques used in occupational hygiene and occupational medicine are well-developed), biological entities, bacteria and viruses, require quite different techniques.

Classification of Micro-organisms on the Basis of Hazard

Clinical microbiologists can never know with any degree of certainty what is in the specimens they receive. Nor, in many cases can pharmaceutical research and development departments that receive blood or other specimens from human volunteers or animals in the course of their investigations into new products know exactly what hazardous micro-organisms may lurk in such materials. For these workers, there are several systems that classify micro-organisms (viruses, rickettsias, bacteria, microfungi, protozoa, and helminths) into groups or classes on the basis of the hazards they offer to those who work with them. This concept was first introduced in the United

TABLE VI *Hazard groups for micro-organisms**

Group	Category
1	An organism that is most unlikely to cause human disease.
2	An organism that may cause human disease and which might be a hazard to laboratory workers but is unlikely to spread in the community. Laboratory exposure rarely produces infection and effective prophylaxis or effective treatment usually are available.
3	An organism that may cause severe human disease and present a serious hazard to laboratory workers. It may present a risk of spread in the community but usually there is effective prophylaxis or treatment available.
4	An organism that causes severe human disease and is a serious hazard to laboratory workers. It may present a high risk of spread in the community and usually there is no effective prophylaxis or treatment.

*From a report by the Advisory Committee on Dangerous Pathogens[3]

States of America some twenty years ago and developed further by the World Health Organization (WHO),[1] the United States Public Health Service,[2] in the United Kingdom by the Advisory Committee on Dangerous Pathogens (ACDP)[3]—and modified by both the European Federation of Biotechnology (EFB)[4] and the Organization for Economic Co-operation and Development (OECD).[5] Several other countries and organizations have introduced their own systems but in all of them there are four hazard or risk groups. That of the ACDP, and applied by the Executive in the UK, is shown in Table VI. The ACDP publishes lists,[3] revised from time to time, of micro-organisms in its Hazard Groups 2, 3, and 4; organisms not in these lists are considered to be in Group 1.

Categories of Containment

The word 'containment' is used to encompass all the precautions that are deemed to be necessary to prevent, in so far as possible, the escape of micro-organisms from their containers into the environment and to reduce the possibility that they might cause disease in man and other animals. It includes equipment, technical methods, mechanical and structural facilities, security and isolation, and decontamination procedures. A separate set of such precautions is provided by the ACDP for each of the four 'Hazard Groups': simple and relatively undemanding for micro-organisms in Group 1 and becoming stricter for those in the three higher groups. The 'Containment Categories' are numbered 1–4 in parallel with the Hazard Groups. The precise requirements are too detailed to be given here and reference should be made to the ACDP Report.[3]

In Britain separate but similar sets of precautions for work with genetically-manipulated micro-organisms are prescribed by the Advisory Committee on Genetic Manipulation (ACGM).[6]

These systems work well in the laboratories for which they were designed but are not really satisfactory for industry where, unlike clinical laboratories, very large volumes of bacterial and viral cultures and quite different techniques often are used. Problems may arise also in laboratories and industries where both natural and genetically-manipulated micro-organisms are handled, and where slightly different precautions for each kind are required by the ACDP and the ACGM. It may well be that at some future date the precautions will be modified for industry along the lines of the systems formulated by the EFB[4] and the OECD,[5] which are intended to apply to both kinds of micro-organism.

Identifying 'Hazardous Substances'

Micro-organisms are not like chemicals. The amount of any chemical present in the environment can be measured accurately; such amounts are finite and their effects are well-documented. Micro-organisms, on the other hand, are difficult and (in some cases) impossible to enumerate. They may die rapidly or increase dramatically in numbers; their virulence (ability to cause disease) is variable, even within species, depending on factors such as the numbers encountered by their victims, on the route of infection—whether it be by inhalation, ingestion, or accidental inoculation—and on the natural or acquired resistance of the host to infection.

Most of the organisms used in industry are in Group 1, defined by the ACDP as 'unlikely to cause human disease'. Some may be in Group 2, which 'may cause human disease'. Pharmaceutical companies that manufacture vaccines and other biologicals may use some of those in Group 3, which 'may cause serious human disease', but there are no industrial applications for the really dangerous pathogens in Group 4.

Assessment and Evaluation of Risk

Fortunately, most industrial microbiologists know the names and properties of the organisms that are used in their laboratories and factories. One must begin, of course, by ascertaining the hazard group to which the organism has been assigned:

Group 1

Care must be taken not to dismiss organisms in Group 1 as 'harmless' and therefore outside the Regulations. Exposure to very large amounts of 'harmless' organisms may be hazardous. For example, some fungi in Group 1 are used to evaluate the resistance to spoilage of manufactured goods such as fabrics: inhalation of the spores of such organisms may induce allergy.

Allergic responses to the enzymes or other metabolic products of some Group 1 bacteria—*e.g. Bacillus* spp.—are not uncommon.

Group 2

Clinical microbiologists know that many of the organisms listed in Group 2 are found in quite large numbers in soil, water, foodstuffs, and in any kitchen. However, as under certain circumstances such organisms can cause human infection, they are placed in this group and therefore must be included in the assessment—albeit as presenting 'minimal risk'.

Group 3

Group 3 micro-organisms are the agents of most concern, both in the laboratory and in industry. They will figure largely in an assessment programme.

Laboratory and Industrial Infections Associated with the Organisms

What the assessor needs to know is whether a given organism is known to be troublesome—in how many job-associated infections has it been implicated? One book does give this information[7] although it is mostly concerned with infections acquired in laboratories. Two others[8,9] describe the microbiological hazards of occupations. The lists of Prescribed Industrial Diseases[10] also are a useful source.

In assessing the risk offered by a particular micro-organism one must consider not only its intrinsic ability to cause an infection—that is, in which Hazard Group it has been placed—but whether it can do so under the circumstances in which it is used in the laboratory and production facilities ('open' or 'closed'), the volumes handled and therefore the extent of exposure under normal conditions, whether accidents ('escapes') could occur, and whether such escapes would be likely to be in the form of aerosols (which may offer a hazard greater then mere spillages). Generally it may be assumed that the individuals who handle or process the organisms are in good health, but there is always the possibility that some of them may be immuno-compromised or are receiving medication that could lower their natural resistance to infection.

Volumes and Quantities of Cultures and Materials

As indicated above, quantitation generally is difficult. But although the same volumes of different cultures obviously will contain different numbers of cells, the actual amount used or processed in any one batch or vessel is very important in the assessment of risk. A step-by-step evaluation of a laboratory procedure or an industrial process, from original inoculum or seed culture to final product (or to the point where the agent is killed) will reveal any areas where workers may be exposed to large numbers. Particular attention should be given to the final treatment and the disposal of infected or possibly infected waste.

Who Should Make the Assessment?

Assessments of microbiological hazards should be made by experienced microbiologists who are well versed in the official and other publications on safety in microbiology and biotechnology. Individuals from other fields or laboratory disciplines are unlikely to do assessments sufficiently well.

Unfortunately, although there are training courses in risk assessment for chemists and engineers, and guidance is given by the Royal Society of Chemistry[11] there are as yet neither courses nor official information for microbiologists who are faced with the problems of COSHH assessments. There is also an understandable reluctance on the part of some companies to employ external assessors, especially if they have already microbiologists on their staff. Smaller firms, however, would be well advised to seek the assistance of professional microbiologists—such as those in university departments or public health laboratories. In many laboratories and industries a full assessment can be made only by chemists, microbiologists, and engineers working in concert.

References

1. 'Laboratory Biosafety Manual', World Health Organization, Geneva, 1983.
2. 'Biosafety in Microbiological and Biomedical Laboratories', U.S. Government Printing Office, Washington D.C., 1988.
3. 'Categorization of Pathogens According to Hazard and Categories of Containment', HMSO, London, 1984.
4. W. Frommer *et al.*, *Appl. Microbiol. Biotechnol.*, 1989, **30**, pp 541–552.
5. 'Recombinant DNA Safety Considerations', Organization for Economic Co-operation and Development, Paris, 1986.
6. 'Health and Safety (Genetic Manipulation) Regulations', HMSO, London, 1989.
7. C. H. Collins, 'Laboratory Acquired Infections: History, Incidence, Causes and Prevention', Butterworths, London, 1988.
8. J. M. Harrison and F. S. Gill, 'Occupational Health', Blackwell, London, 1987.
9. C. H. Collins and J. M. Grange, 'The Microbiological Hazards of Occupations', Science Reviews, Leeds, 1990.
10. 'Social Security (Industrial Injuries) (Prescribed Diseases) Regulations (SI 1985 No. 976) and Subsequent Amendment Regulations 1986 and 1987', HMSO, London, 1988.
11. 'COSHH in Laboratories', The Royal Society of Chemistry, London, 1989.

CHAPTER 8

Compliance in the Newspaper Printing Industry

S. KING

Introduction

News International produces approximately one third of the country's national daily and Sunday newspapers. The principal production site is at Wapping, London, where the staff of over 3000 can be categorized into three basic groups: production, journalism, and others. Each of these groups consists of around a thousand employees.

It is recognized that there are responsibilities to non-production staff but so far as exposure to hazardous substances is concerned these largely can be discounted. In the context of COSHH, most attention must be centred on the production environment.

Previously, newspaper production was based in and around Fleet Street, in cramped and dirty conditions. The move to the new London Docklands changed radically the surroundings and working methods. Expectations and requirements of the production process have been altered, bringing opportunities to enhance standards of safety and the provision for occupational health.

The processes have changed technically in a variety of ways: heavy printing plates are used no longer, thus eliminating hazards from lead and nitric acid. On the other hand, diverse photographic chemicals have been introduced—a matter of particular relevance in the period of transition to producing newspapers in colour.

How a Newspaper is Produced

Nowadays, journalists compile and edit their stories directly, using their own computer terminals. When material is in a form suitable for publication it is sent virtually instantly, at the touch of a button, to the Photocomposing Department. Typesetters are not needed.

In the Photocomposing area, the stories are transformed into 'bromides' in preparation for making up pages. Then, each completed page is photographed, full-size, by means of a large camera and the film developed using conventional dark room equipment and processes.

Next, the page-size film negative is transferred to the Platemaking section. Currently, the method of printing is letterpress, using steel printing plates with polymeric coating. The coatings are sensitive to ultra-violet light and cure when exposed to it for a prescribed period of time. The negative is placed over the polymer side of the plate and exposed: the clear areas of film permit penetration of the light, while dark areas block this: after exposure, a positive is formed in the polymeric coating—areas subjected to the ultra-violet light having cured and the uncured portions being washed away easily in a machine designed specially for the purpose.

The finished plate has a raised surface comprising areas of cured polymer derived precisely from the negative, and ultimately this will be used to transfer printing ink to the paper.

The top edge of the printing plate is bent over to form a lip. On the plate cylinder of the press there is a narrow horizontal channel into which the lip of the plate is hooked. There are magnets flush with the surface of the cylinder and these hold the steel backing of the plate in place. Since the plate is very flexible, it is smoothed into position on the cylinder carefully.

A continuous web of paper is passed through the printing press; the cylinder rotates and half-way through its cycle it collects ink, which is pressed down on the newsprint as the cycle is completed. Each web of paper passes through two printing units, so that both sides are printed, then the webs are collated, folded and cut automatically to produce the finished copies. The number of pages in each newspaper can vary from day to day, and the machines can print tabloids such as *The Sun*, or broadsheets such as *The Times*.

The newspapers are counted mechanically and stacked before being tied and passed in the form of parcels to the Distribution area. There they travel along conveyors to the delivery vehicles: in the course of a night's production, more than one hundred lorries and vans will be loaded.

Hazardous Substances

Table VII lists a number of substances that might present risks in the course of the production processes described above.

In the first instance it is necessary to compile a complete inventory of all

TABLE VII *Substances used or arising in the production of newspapers and associated activities*

Production process	*Substances used or arising*
Photographic processing	Developers, fixers, de-scalers
Plate making	Polymers sensitive to ultra-violet light
Printing	Printing ink, ink mist, paper dust
Cleaning	Paraffin, proprietary cleaners
Loading of lorries	Motor exhaust fumes

the substances that are used or produced at the unit, with a view to identifying any that may be hazardous.

However, the legislation addresses the *control* of hazards, not necessarily their elimination: it requires that risk presented by the use of a substance be assessed. The substance may be 'hazardous' in the sense that it could cause harm to health so that assessment is needed to determine the degree of risk in the actual conditions of use.

In the light of this approach, even hazardous substances may be used provided the risk they present under actual conditions is assessed to be acceptable.

Epidemiology

When reviewing research data relating to epidemiology the immense change in the technology of newspaper production should be kept in mind. The data available currently all relate to former methods of production, and it should be remembered too that many people were employed in the trade on a casual basis and did not work regularly in any one particular printing house.

Because of this the results of studies of earlier records are likely always to be inconclusive, neither demonstrating the presence nor the absence of a link between newspaper printing and adverse effects on health. Leon considered data for mortality of male members of two printing trade unions at Manchester over the period 1950 to 1983 (an inquiry instigated to review a 'cluster' of deaths from cancer of the bladder) but results were not conclusive and no consistent trend was noted.[1]

In a study of cancer of the lung among newspaper machine men, Leon *et al.*[2] concluded that this was the most important single site for cancer in the population concerned but were not able to attribute the disease directly to occupation.

An independent review of the surveys of the Manchester data was commissioned by the Newspaper Publishers' Association but it is thought likely that the outcome will be of academic interest only. The changes which have taken place in working conditions in recent years put the figures largely in the context of history.

The Newspaper Publishers' Association is organizing also a survey of morbidity in the industry: the causes of death among pensioners across the whole spectrum of employment within newspapers will be collected and analysed.

A pilot study for this was conducted by Reynolds,[3] using records of people who were employed by St. Clement's Press and who died between 1980 and 1988. Complications noted when assessing the data were:

(i) differing patterns of employment
(ii) membership of the pension schemes was voluntary, so there was loss of contact later with non-members
(iii) frequently the death certificates gave only vague details of occupations.

Because of the limited scale of the survey and complications such as those mentioned above it was not found possible to draw conclusions.

Ink

Ink of course is a key substance for assessment in newspaper printing. Many varieties are in common use and the particular type considered here is a rotary letterpress black ink with a mineral oil base. (Other types of ink, including water-based, are used in other methods of printing.)

The main components of this newspaper ink are carbon black and mineral oil; it includes also small quantities of clay fillers. The oil base is described as 'Category 3 aromatic oil'—and causes concern because of possible carcinogenic effects. However, the data available are inadequate for more precise assessment at this stage.

Table VIII, page 104, shows routes of entry, hazards, and risks associated with inks of this type. With rotary presses there may be 'ink fly'; if good hygiene practices are not observed there may be contamination of the skin and resulting problems; during cleaning in particular, there may be eye contamination. Ingestion of ink is uncommon. Further details follow:

(i) Skin Contact

The aromatic mineral oils in the ink are associated with dermatitis or, more seriously, with oil acne and even cancer. It is thought the serious conditions are unlikely except in extreme instances of personal neglect and poor hygiene. Regular washing with soap and water is to be encouraged; prolonged and repeated contact, without suitable protection, should be avoided. Overalls and items of protective clothing should be cleaned regularly. Contaminated rags should not be stuffed into pockets. An occurrence of contact dermatitis can give an indication of problems and should be countered by placing particular emphasis on good housekeeping, the wearing of suitable gloves, and on renewed training in safe systems for use of the material.

TABLE VIII *Routes of entry, hazards, and risks in printing with ink based on mineral oil*

Printing activity	*Route of entry*	*Hazard*	*Risk*
—	Ingestion	low	low
As Ink Fly during printing	Inhalation	medium	medium
Cleaning or maintenance	Eye contamination	low	high
Cleaning and printing	Skin contact	low	high

(ii) Inhalation

It is recognized that rotary presses are likely to generate 'ink mist'—a fine aerosol that is thrown off as a press turns at high speed. The mist in turn combines with paper dust to produce what is known as 'ink fly'. Inhalation over the long term of high concentrations of ink fly cannot be discounted as a health risk. A fuller explanation of ink fly, and of its control, appears later in this chapter.

(iii) Eye Contamination

If the eyes are not protected when presses are being cleaned there is a danger of ink splashing into them. (The ink may be diluted with the solvent cleaners used, and probably also will contain paper dust.) In such cases there should be prompt irrigation of the affected eye and medical advice should be sought if pain persists. If treated promptly in this way it is unusual for splashes of ink in the eye to cause more than short-term irritation.

(iv) Ingestion

It is hard to envisage circumstances in which the ingestion of significant amounts of the ink could occur accidentally. Data provided by suppliers suggest that if ingested in moderate amounts the ink would be harmless: larger quantities would cause irritation.

Assessment of 'Ink Fly'

'Ink fly'—comprising fine droplets of ink mixed with particles of paper dust in an aerosol—is generated at the rollers of newspaper printing presses.[4] Factors governing the amounts produced include:
(i) the type and composition of ink used
(ii) the speed of rotation of the machine
(iii) the degree of accuracy with which rollers are adjusted.
In general, the faster the running speed the greater the amount of ink fly.

As mentioned earlier, it remains uncertain precisely what the effects of prolonged inhalation might be but common sense dictates that controls should be applied to reduce so far as possible the production of ink fly and to minimize exposure to it.

Category 3 mineral oils may contain, along with other polycyclic aromatic hydrocarbons ('PAHs'), measurable amounts of benz-alpha-pyrene; it may be that the carcinogenicity of some mineral oils is related to their content of PAHs, but research on this point is far from conclusive.

Organization for Assessments

Guidance notes on the Regulations refer to assessment by 'a competent person'. Competence in this connection is not defined specifically and it is for the employer to decide what level of experience and qualification represents an appropriate degree of competence for the work in hand.[5]

Except in the most straightforward of cases it is unlikely that any single person would be capable of carrying out all aspects of an assessment; in most instances, a team approach is appropriate. In order to achieve an effective programme one requires a well-balanced and motivated team, with good leadership; there should be within the team representatives of all relevant spheres of activity at the unit.

Team Formation

In the instance under review, the structure of the organization was considered and employees with appropriate skills and knowledge were invited to take part in a steering committee. This group comprised:

Manager (Occupational Health and Safety),
Purchasing Executive,
Manager (Production Stores),
Technical Services Supervision,
Employee Safety Representative.

Each member of the committee was in a position to call upon resources from his own department when need for this arose, and specialist advisers could be consulted as necessary.

The responsibilities of the committee were shared among members but each of the members had also specific functions, as follows:

Manager, Occupational Health and Safety

(*a*) To co-ordinate group activities and to set targets to ensure that the legislative time-table was met
(*b*) To provide advice on the requirements of the Regulations and to interpret them in the context of the activities of News International
(*c*) To maintain contact with other health and safety practitioners, particularly within the same industry, thus enabling an exchange of information.

Purchasing Executive

(*a*) To liaise with suppliers to ensure that data on safety was provided for all the materials brought in
(*b*) To ensure that when requisitions were raised, the department concerned would be reminded to obtain and to make use of the relevant information.

Manager, Production Stores

(*a*) To review the types and quantities of materials used currently and to predict demand for them in the future, with a view to ensuring that stocks of substances hazardous to health be kept to a minimum and bearing in mind that any unwanted materials would have to be returned to the suppliers concerned or sent for disposal
(*b*) A further function in this instance was to ensure adequate supplies of personal protective equipment, and to control the issue of such equipment.

Technical Services Supervisor

(*a*) To advise on substances used within the maintenance and engineering workshops
(*b*) To make the group aware of any modifications planned which might affect safe working systems or interfere with control measures.

Employee Safety Representative

(*a*) To maintain contact with other safety representatives and enable the company to fulfil its statutory duty to advise and inform
(*b*) To gather for consideration any comments about systems of working where employees perceived hazards and offered suggestions for improvements
(*c*) To bring to attention any deviations from established procedures, and to explain reasons for them
(*d*) To gauge actual use of personal protective equipment (as distinct from figures for issues of such equipment)
(*e*) To sound out employees on proposals from the group, in case they might be thought impracticable.

Activities of the Steering Committee

The committee met for the first time early in March 1989 and set a timetable for implementation by 1.10.89. A general circular to introduce the group and to explain its objectives was produced and sent to all heads of departments within the company.

With the help of the purchasing member of the committee, a standard letter was sent to every organization on the company's list of suppliers: this referred to existing obligations under the Health and Safety at Work Act, as amended, and drew attention to the new Regulations and their implications. It requested that all suppliers send current product safety data sheets for each of the materials that were purchased from them.

When the safety data sheets were received and examined it was found that in several instances information was inaccurate or misleading. (It was felt that in some of these cases a purchaser might not have realized that incorrect information was being supplied.) Problems of this nature may be reviewed in general terms as follows:

(i) Hazards Over-Stated

Should hazards be over-stated in a data sheet there is a danger that customers will be reluctant to use the product and the benefits of sales will be lost. In one instance we received a data sheet for a type of photographic developer which seemed to require unusual precautions and to state exaggerated effects from exposure (it included a warning that 'brain damage' might result from repeated exposure). The information was questioned with the supplier (a distributor in Britain of the chemical concerned) and it transpired later that by mistake information relating to the manufacture (not the use) of the substance had been given. Had this not been brought to the attention of the supplier no doubt he would have continued to send out the incorrect details.

(ii) Hazards Under-Stated

Sometimes, data may be presented in such a way as to under-state or even conceal hazards. (Alternatively, it happens that hazards which might appear insignificant come to assume much greater importance at a later date.)

Examples have come to light of incorrect advice as to precautions. As an instance of this, a sheet for an aromatic hydrocarbon blend used as a cleaning material for the printing rollers gave emergency action for skin contamination as to apply barrier cream. (The use of barrier cream, as its name suggests, is recommended to protect the skin in certain situations but *before* contact with the substance takes place; adding barrier cream to the skin after such contact would serve only to maintain it. It was interesting to see printed at the foot of the sheet concerned a 'disclaimer' in which it was said that the supplier accepted no responsibility for the use of, or reliance upon, the safety information given.)

Every department within the company was required to compile a list of all the substances in use currently (even if they were not perceived as being hazardous). The lists were returned to a central office, an index of substances was created—and as the safety data sheets were received they were checked against this. Companies not responding within a certain period of time were sent reminder letters by the Purchasing Department.

As a general comment, the more satisfactory response was received from larger suppliers; they were familiar with the Regulations and aware of responsibilities under them. Some small suppliers, on the other hand, seemed oblivious to the Regulations and, indeed, to other health and safety legislation. In a few cases, as indicated above, safety data sheets had been prepared and issued but with insufficient understanding of why this was being done. Sometimes data were inaccurate, and often incomplete. In cases such as these the suppliers were sent copies of guidance information prepared for those trading with the printing industry.[6]

The steering committee considered compiling pro forma data sheets with a view to overcoming errors and omissions such as those indicated but it was decided not to pursue this idea. The majority of members felt that it was for the supplier to decide what information should be given. (There was also a possible legal danger that if the sub-committee did not ask for information that had been omitted from a data sheet, and a problem arose later as a result of the omission, the steering committee might be held to be guilty of contributory negligence.)

Internal Assessment

As a data sheet was received it was matched with the department using the substance concerned and sent with an assessment sheet to the manager of that department. The request was made for an initial assessment of the use of the substance and when received this was considered for verification by the group. (Should no assessment be received within a given time, a reminder was sent.)

Each completed assessment received was discussed by the group. If no further action was necessary the assessment was endorsed and a date settled for its review; the assessments were then filed, using the Croner system (which was one of the first to be made available, and appeared to cover our requirements).[7]

It was found that the great majority of the substances in use could be assessed internally with confidence. (Should training be necessary, or protective equipment provided, these things could be arranged.) In some cases it was decided that leaflets would be an adequate means of bringing information for the attention of employees; when training was necessary it encompassed staff at all levels—and it could be, as appropriate, basic instruction or more formal sessions in the classroom.

Some substances could not be assessed properly on an internal basis and two examples of these are given below.

External Assessments

Case Study 1: Measuring Ink Fly

Ink fly usually is measured using the method contained within MDHS 14 'Total inhalable and respirable dust gravimetric'.[8]

It recommends the use of a pump device and a pre-weighed filter. Air is drawn through the filter at a recorded rate of flow and the duration of the sampling logged. From these data and the material collected on the filter it is possible to calculate the volume of air sampled and the concentration of the substances for which sampling is being carried out.

Whenever this is possible, sampling should be for the whole of the normal shift and under normal working conditions. The point of collection of the air should be within the breathing zone of the operator concerned. Usually it is helpful to employ sampling equipment sited on or near specific units in addition to personal samplers worn by the operators. The latter should indicate the levels of exposure for the individual operators while the equipment at selected locations can show whether particular units are at fault, or might cause difficulties in specific conditions or circumstances.

The total amounts of particulate matter gathered by each of the filters are determined gravimetrically and the fraction in each case soluble in cyclohexane ascertained—to indicate the nominal contents of oil.

A maximum time-weighted average figure of 1.5 mg m^{-3} for exposure of an operator over a normal working shift has been recommended. (This relates to the fraction soluble in cyclohexane.) Although it has no legal standing the recommendation is viewed as a realistic limit by the Printing Industry Advisory Committee and is endorsed by the Health and Safety Commission.[9]

In accordance with approved procedure, sampling should be carried out at regular intervals. Provided the supplier and the formulation of the ink remain unchanged, and there are no other changes in the operation of the unit (such as in speed of running, or the arrangements for ventilation), then monitoring at intervals of, say, six months could be adequate. It should be carried out more frequently if operating conditions are changed in any material way.

The monitoring should be in accordance with a suitable written procedure, by staff exercising appropriate care. The filters must be conditioned and weighed in advance, the sampling equipment supervised to ensure the correct rates of flow—and that there is no tampering with the filters—with the subsequent weighing and calculations also accurate and precise. Equipment for the sampling and for the laboratory work may be purchased from several reputable suppliers. Probably it is desirable that the methods used for sampling and analysis should be compiled by a qualified and experienced analytical chemist and supervised in the first stages by such a person (for example, a chartered chemist and Member or Fellow of the Royal Society of Chemistry). The person carrying out the work should be able to refer easily to the qualified person in any case of difficulty at any time. The Royal Society of Chemistry has an 'indicative register' of certified health and safety specialists among its members (in other words, of men and women likely to be able to undertake such work). Membership of other professional societies or institutes can be taken as indicative of appropriate degrees of competence for activities such as sampling and monitoring (as an example, membership

TABLE IX *Typical results of sampling for particulates and cyclohexane-solubles in air in a newspaper press area*

Sample No. and location	*Duration of sampling* (minutes)	*Volume of air sampled* (l)	*Total particulates* (mg m^{-3})	*Soluble in cyclohexane* (mg m^{-3})
1. Personal sample (a)	317	634	3.17	0.49
2. Personal sample (b)	316	632	1.59	0.16
3. Personal sample (c)	302	604	3.45	0.49
4. Folder D5	298	596	3.27	0.41
5. Folder C2	298	596	1.24	0.79
6. Folder 4B	302	604	1.00	0.22
7. Folder 5A	294	588	1.86	0.72
8. Personal sample (d)	296	592	2.84	0.01
9. Personal sample (e)	296	592	0.85	0.51
10. Personal sample (f)	295	590	0.02	0.09

of the British Occupational Hygiene Society). The Department of Trade and Industry offers an 'accreditation' scheme (the National Measurement Accreditation Service, or 'NAMAS') under which, with an annual fee, monitoring systems can be inspected and approved (or otherwise) by officials from the National Physical Laboratory.[10] In a unit in which ink fly is well controlled it may be necessary for monitoring only to complete the first stage of the analysis—that is, the weighing of filters. (Should the weight of a filter have increased by the equivalent of less than 1.5 mg m^{-3} the fraction soluble in cyclohexane is not of concern.) Examples of some typical results are given in Table IX.

Control Strategies. Using the 'hierarchy of preferred solutions' (a classical strategy in occupational hygiene), various steps may be taken to control and to minimize ink fly:

1. Elimination. Realistically it is not possible to eliminate the use of ink for newspaper printing.

2. Substitution. Suppliers of inks should be encouraged so far as possible to continue their development programmes for 'low-mist' inks.

Purchasers of machinery should consider whether to specify a printing system using water-based ink, if this would be compatible with the processes they use. While it is not practical to change machinery with the sole purpose of reducing ink fly this is a consideration to be kept in mind when decisions are made as to the machines offered for sale in the market.

3. Enclosure. Many presses are enclosed already, either individually or in groups. The primary reason for this is to reduce exposure to noise, but a well-designed enclosure can reduce also contact with ink fly.

4. Segregation. A variation of the approach of enclosure that is becoming more popular is to provide enclosures for the operatives. This makes possible the provision of a more tolerable climate and conditions: so far as can be arranged, controls should be within the operator enclosure and the frequency of visits to the machinery area reduced to the minimum absolutely necessary.

5. *Reduced Exposure*. Direct contact with ink mist may occur when an operator has to enter an area to adjust the setting of a machine or to rectify a fault. 'Re-webbing' (that is, re-threading the newsprint through the machine after a break) is a common example. In such cases it is adjacent machinery continuing to run that gives rise to the exposure. Since it is not sensible when paper breaks on one of them to stop all the machines, individual enclosure of machines is the only way in such circumstances that engineering can be used to reduce exposure.

On the other hand, an effective programme of planned maintenance can help reduce the frequency of machinery faults during production. Automatic and remote control systems limit the need for direct human intervention.

6. *Ventilation.* Local exhaust ventilation ('LEV') will improve ambient conditions significantly. Of course it must be well-designed, properly installed, and maintained. Contaminated air should be removed from as close as possible to the source of contamination and transported by suitable ducting to the outside. A filtration system should be incorporated, to make sure that responsibilities to the public at large are not infringed. If local exhaust ventilation is listed as a control measure under the Regulations, schedules of testing and records of results should be kept in accordance with the details stated. (Thorough examination and testing of the equipment at suitable intervals is required: records must be kept for a minimum of five years.)[11]

In cases where it is impracticable to remove the ink fly at source, dilution ventilation should be used. (Such a method is acceptable also if ambient levels are low already.) With this approach, over a period of time, the contaminants will settle and cause discoloration of exposed surfaces.

The system of ventilation should be designed to take account of enclosures, whether of machines or operators. In the operators' area there should be always a positive air pressure so that when access is necessary the clean air forces back the contaminated and not *vice versa*.

7. *Personal Protective Equipment*. The use of personal protective equipment is seen as a last resort. Ideally it should be used only as an interim measure, in anticipation of a more reliable and permanent resolution by engineering means.

However, if 'PPE' is to be used as a control measure for ink fly a basic dust mask is generally all that will be required. The suppliers of such equipment should be consulted for their recommendations. The bulk of the contamination may be found to be paper dust rather than ink mist. Paper dust is classified as a general nuisance dust and as such has a recommended eight hour time-weighted average exposure limit of 10 mg m^{-3} for total inhalable dust. (The limit for respirable dust—that is dust capable of reaching the gas exchange region of the lungs—is half this.) Airborne dust in newspaper printing environments usually is made up of larger particles of paper dust;

however, analytical examination should differentiate between inhalable and respirable fractions.

If concentrations of dust are recorded which exceed the limits shown, it would be considered a 'substance hazardous to health' under Regulation 2.

Case Study 2: Assessment of Cleaning of Press Hall

In the printing machine area a crew carries out regularly full-scale cleaning of the presses. The cleaning shift begins each day at 7.30 a.m. (approximately three hours after the night's production has ended) and involves a number of men working with paraffin and cleaning rags. In the course of the work there is a noticeable smell of paraffin and skin contact with it is common. Medical staff advised that one or two people were complaining of 'fumes' or of rashes and Casella Limited was engaged to carry out monitoring for fumes, using both personal sampling and static sampling devices. A report and recommendations were requested.[12]

Measuring Paraffin Vapours. In the course of the sampling, air was drawn through tubes containing activated charcoal at rates of flow of up to 300 ml/minute. Personal samples were taken within the breathing zones of the operators, with static samples at selected locations in the working area, with a view to obtaining a more general record of the situation.

The samples taken were desorbed and analysed using capillary gas liquid chromatography. Peaks were shown in the range C_9 to C_{11} (bulk paraffin is a hydrocarbon blend, C_{10} to C_{13}, and the airborne samples contained as expected the more volatile components). 'Guidance Note EH 40' gives no occupational exposure limit for paraffin as such but includes white spirit (a similar product) for which the eight hour time-weighted average is 575 mg m^{-3} and the short-term exposure limit (ten minute reference period) is 720 mg m^{-3}.

The highest results obtained from the samples were only one-fifth of the occupational exposure limits for white spirit, suggesting that there should be no significant risk to the health of employees from the inhalation of the vapour during the cleaning operations (Table X, page 113).

However, notwithstanding this satisfactory outcome, a number of factors required attention still. The survey had measured only the possibility of inhalation of the vapour and not contamination of the skin. It was necessary to review the system of work and ensure that the best means practicable were in use: it was found that although exposure to the vapour was well within applicable limits it could be reduced further by introducing relatively inexpensive controls.

Control Strategies. 1. Elimination or Substitution. It is essential that the presses be cleaned each day, to allow for inspection of the machinery and to prevent progressive deterioration in the quality of the products. It would be essential also to use a solvent of some kind in this work, since it is necessary to clean away the ink. There are available water-based cleaning products but

TABLE X *Measurements of airborne paraffin vapour in a newspaper machine room during daily cleaning activities*

Sample No. and location	*Duration of sampling* (minutes)	*Concentration in air* ($mg\ m^{-3}$)	*Time-weighted average concentration** ($mg\ m^{-3}$)
1. Mr A. (Line C1)	170	168	101
2. Mr G. (Line C1)		(sample damaged)	
3. Mr P. (Line A1)	272	34	20
4. Mr D.	273	83	50
5. Static, Machine 1 (C1)	273	10	
6. Static, aisle between Machines 1 and 2 (C1)	274	15	
7. Static near Machine 3 in aisle near door	273	14	
8. Static, near Machine 3 in aisle near door	264	87	
9. Static, near Machine 1	260	7	
10. Static, aisle between Machines 1 and 2 (A1)	261	1	

**Note*
Occupational exposure limits are on the basis of an eight hour working day, five days per week and in making assessments any variance in real working conditions must be accounted for mathematically, so that direct comparisons with the limits may be made. The operators in this instance work from 8.00 a.m. for an eight hour day, four days per week—a total of thirty-two hours.

it has been found that over a period of time these can cause corrosion of the machinery; cleaners of the detergent type are not acceptable either, because of the periods of time necessary for their effective application and drying.

2. Reduced Exposure. When cleaning commenced the presses often were warm still, and since evaporation takes place more readily when heat is applied the warm metal promoted vaporization of the paraffin. It was possible therefore that by revising cleaning schedules the time available for cooling the machinery could be extended, with some reduction in the generation of vapour.

Other possible means of reducing exposure included:

(i) limiting the number of people engaged at any one time in the cleaning area; those not actively engaged in the cleaning could be sent to other work elsewhere

(ii) access to the press areas could be limited until all cleaning was complete

(iii) all cleaning rags and other contaminated matter should be kept in closed bins and removed promptly from the press area.

3. Ventilation. In an instance such as this, where ambient levels were low, dilution ventilation was considered adequate. There might be a need for local exhaust ventilation in certain areas—such as those in which paraffin

was dispensed, or where used material was collected—but in most cases these could open safely to external atmosphere.
4. *Personal Protective Equipment*. All engaged in cleaning duties such as these should wear gloves at all times. The gloves should be of length sufficient to ensure that hands, wrists and arms are protected. Should cleaning rags be dipped into buckets it may be preferable to use gauntlet types, to prevent ingress of solvent over the cuffs.

Suitable overalls should be worn, with fresh changes of clothing at frequent intervals.
5. *Training*. It was necessary to explain thoroughly the necessary principles of good personal hygiene in these circumstances, and (for example) the effect of paraffin of removing fat if not washed away promptly. (The use of a fat replenishment cream can be advantageous in such circumstances.)

Emphasis in training was placed upon establishing a safe system of work and of adhering to it rigidly. In addition, employees were to be encouraged to report to company medical staff or to their own practitioners any signs of illness as soon as they were noted.

Health Surveillance and Keeping of Records

When the assessment records were reviewed it was found that no substance was mentioned for which surveillance was mandatory under the Regulations. Most of the information that was required to be kept for employees involved was being kept already.

However, it was thought prudent to go beyond the minimum requirement under the Regulations and with this in mind a total of twenty production employees was selected at random, each of them to be subjected to an annual medical examination. The screening would be carried out at the company's own Medical Centre and would be concerned particularly with any problems that might be related to substances in use at the plant. The results of the medical examinations could be monitored and this could assist in the early detection of an over-exposure, should this take place. The records arising would be maintained as though this were a direct requirement of the Regulations.[13]

It has been the policy always to submit all production staff to medical examination prior to employment: this policy is continued and is regarded as an excellent way in which problems or susceptibilities already existing may be identified. Employers adopting such a system should note that if it should be known that an individual is susceptible to a particular substance or risk an enhanced duty rests on the employer concerned to ensure the health and safety of the vulnerable person.[14]

Strategy Anticipated for Enforcement

The newspaper printing industry has had for some time a close relationship with the Health and Safety Executive and the link is strengthened through the Printing Industry Advisory Committee and a National Interest Group based in London. Additionally, the employers are members of the Newspaper Publishers' Association, the safety advisers' committee of which meets regularly.[15]

At this early stage is may be possible to make some general observations as to the strategy for enforcement of the Regulations that can be expected. My impression, after discussing the topic with a large number of people in the field of health and safety (including members of the Executive), is that the approach will be rigorous but also reasonable.

There has been a change of emphasis from what might be called in the old days the 'hardware' of safety—the guarding of machinery, and so forth—to the 'software'. The latter term takes account of the intangibles which contribute to a corporate consciousness of safety—including training for safe operating and the systems of work themselves. The publication 'Human Factors in Industrial Safety' provides an illustration of this change of emphasis, as a document of this type probably would not have been considered until recently.[16]

'COSHH' largely is a piece of 'software'—legislation directed towards instruction, training, supervision, and the provision and maintenance of safe systems.

In the early years of enforcement the Executive is likely to concentrate its resources on the larger companies. Should a succession of well-known names be prosecuted the ensuing publicity would help to increase the effectiveness of the Regulations elsewhere. In smaller organizations, at least in the early stages, the main pressure for compliance is likely to be from trade or professional organizations, or as a result of individual difficulties. (The most recent statistics for inspections suggest that on average a company may expect a visit from a factory inspector about every ten years.) The frequency of visits by inspectors suggests that, at least at first, application of the Regulations may be incomplete—with some employers perhaps responding to events that have demonstrated inadequacy and others doing very little.

When approaching a company about COSHH an inspector will expect to find details of assessments. If these exist it is likely that they will be taken at face value. An employer who has taken the trouble to write an assessment is likely to have worked to a reasonable standard and unless there is an obvious flaw the document will be taken to be satisfactory. (The alternative would be for the Executive to duplicate the work which it has neither the time nor the resources to do.)

In Summary

Every industry has its own particular peculiarities and no two assessment programmes will be identical. However, whatever the format, all should follow the same basic principles—companies will be required to recognize, measure, evaluate, and control any substances that could cause harm.

It is important that an assessment be approached in an informed manner and also that the assessors be aware of their own limitations. (No stigma should be attached to a decision to bring in outside assistance.)

Employees, both individually and through recognized union structures, should become involved closely in the efforts of the employer to comply with the requirements of the legislation. Without their contributions and co-operation it is unlikely that control measures can be effective. The real success of a COSHH programme is to be measured not by the paperwork but by what happens in practice.

It is crucial that everyone associated with it appreciates that the assessment and monitoring activities should be continuous. Whenever processes used or systems of work change, the need for new assessments must be considered.

The Regulations provide an opportunity for establishing good programmes in the fields of occupational safety, health, and hygiene—an opportunity that should be taken.

Acknowledgement. The author wishes to acknowledge the assistance of Nicola Coote in preparing this chapter.

References

1. D. Leon, Mortality of male members of two printing trade unions. Unpublished, 1989.
2. D. Leon, P. Thomas and S. Hutchins, Lung cancer in newspaper machine men. Unpublished, 1989.
3. J. Reynolds, 'Pilot morbidity study of Fleet Street production workers', 1988.
4. 'Ink Fly in newspaper pressrooms', HMSO, London, 1984 (ISBN: 0 11 883751 6).
5. 'The Control of Substances Hazardous to Health Regulations 1988, General Approved Code of Practice on Control of Substances Hazardous to Health', HMSO, London (ISBN: 0 11 885468 2).
6. 'The provision of health and safety information by manufacturers, importers and suppliers of chemical products to the printing industry', HMSO, London, 1986 (ISBN: 0 11 883852 0).
7. 'Record Keeping Book for COSHH', Croner Publications, New Malden, 1989.
8. 'Methods for Determination of Hazardous Substances. 14 Total inhalable and respirable dust gravimetric', HMSO, London.
9. Printing Industry Advisory Committee. (Advisory Committees are appointed by the Health and Safety Commission under Section 13(1)(d) of the Health and Safety at Work Act 1974.)
10. National Measurement Accreditation Service. ('NAMAS' allots a registration number to each organization concerned, with details of the activities it has accredited.)
11. 'Health and Safety The Control of Substances Hazardous to Health Regulations 1988. Statutory Instruments No. 1657, 1988', HMSO, London.
12. Casella Limited, Bridge Wharf, Caledonian Road, London.
13 S. Burnouf, Annual Health Screening of printing production workers. Unpublished, 1989.

14. S. Burnouf, Pre-employment health considerations for newspaper production workers. Unpublished, 1988.
15. Newspaper Publishers' Association, 34 Southwark Bridge Road, London.
16. 'Human factors in industrial safety', HMSO, London (ISBN: 0 11 885486 0).

CHAPTER 9

Application in Retail Business

Compiled by the Editors

Introduction

Just as, numerically, the small unit remains a very important component of manufacturing industry, so also is retail business in Britain. Unlike in the United States, small shopkeepers here were not protected by legislation against the buying power of the multiples, and inevitably their higher prices to the public contributed to driving thousands of them out of business. Nevertheless, as Table XI shows, significant numbers exist still—in excess of 200 000, employing some three quarters of a million people (out of a total employment in retail business of about 2.3 millions).

Retailers with single shops are classified in the official figures as shown in Table XII, page 120. Briefly stated, food retailing accounts for the largest group of single-outlet retailers (31.2%) and mixed business and hire and repair (2.0% and 0.8%, respectively) the smallest.

When the large and small multiples and the co-operative societies are included, food retailing still dominates (with 98 000 outlets), 'household goods', and 'clothing footwear and leather goods' second and third most important in terms of numbers (60 000 and 58 000 outlets, respectively).

The diversity of shops and types of goods stocked is much wider than the classifications in the official statistics suggest. While food shops of all kinds may be the most familiar (ranging from the smart delicatessen to the 'farm shop' at the roadside) they include booksellers, dealers, domestic, and other electrical goods (computers, telephones, and so forth), fancy goods, hair-dressers, jewellers, paper suppliers, painting and decorating emporiums, pet-shops, pharmacies, photographers, newsagents, stationers, second-hand shops—and no doubt many others. The ranges of goods in shops of all types

TABLE XI *Numbers of retail businesses and employment, 1987*

Type of business	*Total number*	*Outlets* (number)	*Persons engaged* (thousand)
Total retail trade	240,853	345,467	2,319
Single outlets	213,378	213,378	786
Small multiples	26,613	69,384	306
Large multiples	862	62,706	1,228
Of which, co-operative societies	90	4,691	100
Food retailers	73,681	98,016	806
Single outlets	66,504	66,504	224
Small multiples	6,964	18,538	83
Large multiples	213	12,975	499
Drink, confectionery, and tobacco	47,296	59,810	272
Single outlets	45,241	45,241	177
Small multiples	1,976	4,797	22
Large multiples	79	9,771	73
Clothing, footwear, and leather goods	31,162	58,380	293
Single outlets	24,959	24,959	80
Small multiples	5,970	16,299	55
Large multiples	233	17,122	159
Household goods	42,760	60,406	307
Single outlets	37,253	37,253	126
Small multiples	5,393	13,603	58
Large multiples	113	9,549	123
Other non-food retailers	38,973	52,473	233
Single outlets	33,333	33,333	127
Small multiples	5,495	13,819	60
Large multiples	145	5,321	47
Mixed businesses	4,937	11,363	375
Single outlets	4,225	4,225	47
Small multiples	650	1,866	27
Large multiples	62	5,272	302
Hire and repair businesses	2,045	5,020	32
Single outlets	1,863	1,863	5
Small multiples	165	462	2
Large multiples	17	2,695	25

change as time goes on—reflecting changes in costs and supplies, technical progress, alterations in taste, and in the preferences of customers. The introduction of new materials can mean that items seemingly familiar in fact are made from different things. The methods of selling and of packing for sale change too. From time to time, entirely new types of shop emerge—such as for retailing brass door-knobs, electric light fittings, computer programmes, model aircraft, and entertainment films in video form for hire or sale.

TABLE XII *Types and numbers of retailers with single outlets only*

Type of retailer		*Outlets*	
		Number (thousands)	*Per cent of total*
Food		66.5	31.2
Drink, confectionery, and tobacco		45.2	21.2
Clothing, footwear, and leather goods		25.0	11.7
Household goods		37.3	17.5
Other non-food retailers		33.3	15.6
Mixed businesses		4.2	2.0
Hire and repair businesses		1.9	0.8
	Total	213.4	100.0

Common Problems

It was pointed out elsewhere in this book that it was not at all unusual for a large manufacturing establishment to use tens of thousands of different substances, and that for many of these under the Regulations assessments had to be made. In retailing too, very large numbers of different substances are to be found (although admittedly, in most cases, each in relatively small quantities). Notwithstanding the consolidations of recent years, retail business remains very diverse—and the list of substances on the shelves and in storage bays at any one establishment, even in the same trade, quite often would differ greatly for a list compiled elsewhere.

In some trades the proprietor or manager will have good knowledge in detail of the types of goods for which he is responsible—this would be true, say, of the pharmacist, suppliers of films, paints, dopes, wines, *etc.*—but in large numbers of retail establishments in a variety of categories the proprietors or managers could be expected to have only limited knowledge in detail of the composition of the many things they offer for sale. In retail trade, branded marketing reaches its apogee: in a sense, even the 'own name' retailers sell branded products and must be at least as responsive to changes in relative costs of raw materials as are the leading manufacturers in fields in which they compete—like foods and pharmaceuticals. Thus, if a shop manager does happen to know in January of a particular year the compositions of the items on his shelves it is all too likely that by March at least the formulations of many of them will have been changed.

There is an important difference in this respect between the large multiples and the individual outlets: while the former will be able to keep extensive record systems and carry their costs, and can require suppliers to advise them of changes in formulations, the small man (or woman) has no comparable resources. For many multiple organizations the implementation of the Regulations will be much as with large organizations in other fields (essentially, the collection and assimilation of large volumes of data), but the limit of what is possible for the small retailer determines that he must find effective and acceptable ways of simplifying the task.

While retailing may be similar in many respects to other types of business—commercial and industrial—it is for the most part rather different in the numbers and diversity of members of the public with which it must have regular contacts. Many types of enterprise receive the public; bus lines, railways, offices, and others sell tickets and accommodate customers, but usually in rooms such as booking halls set aside for the purpose; typically, in the commercial world, visitors will be seen at reception desks and conducted to a meeting room or perhaps to the private office of the person to be seen. Some factories have facilities for regular tours, and others (even nuclear establishments) have occasional 'Open Days' when employees and their families are welcomed inside. However, none of these examples is quite like the retail shop, where, each day, large numbers of members of the public enter, walk round, and even handle the goods on display. During business hours, most retailers are truly 'open to the public'. (Very few are in the position of the eccentric bookseller who would turn way anyone he suspected only of wishing to browse.) Most people can be expected to be perfectly reasonable and to behave responsibly when surrounded by the goods on show, but 'the public' includes children, mendicants, and not a few eccentrics of all ages. (Nowadays it includes even a small number of people intent on blackmail, or on making trouble in other ways.) Apart from some problems with the public at large which perhaps might be foreseeable, there are others—arising say from drink or from illness—which are quite difficult to anticipate.

Problems of access will differ in degree from establishment to establishment; in small shops—particularly if converted to 'self-service'—it is not unusual to see a clutter of steel baskets with assortments of special offers and the like which may make it awkward even for a single shopper to move about freely. Usually the multiples have more space available but even then there is need for alertness in the placing of displays and offers, and an ability to correct mistakes quickly when these occur.

There should be a well-established emergency drill from the removal of damaged or smashed containers, and for making surrounding areas safe (including disinfection). Similarly, designated members of staff should be ready at any time to provide simple emergency treatments (such as for cuts).

There are still many shops in old properties, in basements or on two or more floors, and in adjoining buildings linked by new doorways or the removal of party walls. Often in such properties the sales departments on different levels are accessible by fixed steps or stairways, which sometimes can be quite narrow or winding. Such stairways always should be well-lit and not obstructed by goods. If appropriate (when, say, there is not space for two people to pass side-by-side), signs stating priority should be displayed.

In our experience, lifts of the 'Paternoster' type (where the cars have no doors and, like prayers, ascend constantly) are found only in public buildings of a certain vintage: the 'people-movers' in retail shops and malls take the more familiar forms of conventional lifts, escalators, and moving walkways. Here again, there could be some problems in connection with safety and the transport of goods; shopkeepers and managers are advised to warn customers as to the proper use of such equipment, including controlling the baskets and parcels, and supervision of children. Sometimes (again, particularly in older buildings) lifts and escalators may be small and of rather restricted capacity, and this should be made clear by warning signs. Low ceilings, projecting corners and the like should be marked and padded if possible.

In the past a grocer always would keep a cat; today, as a rule, animals are not allowed in food shops, but this is not always the case with other types of retail establishment. The retailer must be conscious of possible dangers from an occasional inquisitive dog or cat—or perhaps even from some more exotic pet brought in and let loose accidentally.

Apart from allowing for the inevitable contact with the public in all its forms another important problem in common among many retailers, both large and small, is the possibility of administering a dose by mouth. Usually, in manufacturing industry this would be an unlikely route for exposure to a hazardous substance but many retailers sell food and drink for consumption at home, for 'take aways', or in restaurants. In addition, qualified pharmacists supplement the work of medical practitioners and offer treatments for ailments—normally doses given by mouth, although also by injection. There would appear therefore to be for retail businesses especially an important area in which the requirements of COSHH (supplementing other legislation) may be rather different from others considered so far.

Various organizations in the retail trade have issued notes and guidelines related to specific types of question—for example, the National Pharmaceutical Association 'Information Leaflet No. 9', which provides a summary of the Regulations and a series of appendices on points to be considered, some of which are quoted here (with permission) as Figures 10–14 (pages 123 to 127), inclusive.

The suggestions made in this review, which it is hoped will be of help to the retail businessman or manager, may be summarized as follows:

(i) Applying the letter of the Regulations as fully as practicable in his own circumstances

(ii) Making a critical survey of arrangements and practices in the outlet or outlets concerned, and correcting any hazards noted.

Applying the Regulations

For the large or small multiple organization the advice must be akin to that given in earlier chapters—to consider buying policy and to obtain data on the composition of every product brought into the premises and sold. Having in this way compiled a comprehensive list of substances, the risks in practice in each case should be assessed. It is important to remember that the assessments should include any risks associated with ancillary materials—such as glass bottles, plastic films, waxes, or synthetic coatings. The buying

Points to consider:

IN THE DISPENSARY

- substances covered by the CPL regulations (solvents, pesticides, poisons, other chemicals)

  your list may include:

acetic acid	hydrogen peroxide
ammonia	isopropyl alcohol
carbon tetrachloride	methylated spirits
caustic soda	oxalic acid
chloral hydrate	phenol
chloralose	potassium permanganate
chloroform	silver nitrate
ephedrine and salts (in powder)	sodium chlorate
ethanol	sodium hypochlorite solution
ethyl chloride	strychnine
formaldehyde	sulphuric acid
hydrochloric acid	

NB: This list is not comprehensive. Many pharmacies will stock items other than these (You may find that you stock fewer than these.)

- micro-organisms present in body fluids (blood in screening tests or from sharps waste).
- any medicines which might present a risk in **handling**

  eg — some phenothiazines may cause skin problems.

  — cytotoxic drugs should not be handled

  (in any case many medicines are supplied in blister packs which reduce exposure).
- any chemicals (medicinal or non-medicinal) which are in powder form and could present a risk by inhalation.

NB: Remember that **medicines**, etc. are only covered by the COSHH regulations if they present a risk in the way they are used in the workplace. The medical effects of the medicines on patients are not relevant to these regulations.

IN THE "SALES AREA"

- you may stock and sell substances which are potentially hazardous such as cleaning fluids, bleaches, weedkillers, garden pesticides, nail varnish remover, solvents and adhesives etc. These will be sold in closed containers and only present a hazard in the event of breakage and/or spillage.

ELSEWHERE

- cleaning fluids and disinfectants used by regular staff or by cleaners who work outside shop hours.

It will probably be sufficient to make a simple assessment of areas other than the dispensary, listing the potential hazards and stating provision for breakage and spillage.

When considering what to do in the event of breakage and spillage you should bear in mind the following:

- ▶ Spillage powders are available, such as Encap, which effectively "mop up" liquid spillages. There are two types—one is chlorinated and therefore useful for blood or other body fluids, and the other non-chlorinated and useful for chemicals which might react with chlorine causing more problems than the spillage itself! Details of suppliers can be obtained from the NPA Information Department.
- ▶ Powders should be removed using a vacuum cleaner.
- ▶ Wear protective clothing (gloves) if the substances are likely to be harmful if handled. This is particularly important in dealing with sharps waste or other sources of contaminated blood which could cause HIV or Hepatitis B infection.
- ▶ First aid procedures for dealing with skin or eye contact. Details of suppliers of eye-wash bottles are available from the NPA Information Department.

You must make your own assessments according to what you stock and how you use it, and what else you do and how you do it.

FIGURE 10 *Points to consider in retailing, as summarized by the National Pharmaceutical Association*

Sample Assessment

The Control of Substances Hazardous to Health Regulations, 1988

ACETIC ACID, GLACIAL BP

1 **Nature of Hazard**
 i Corrosive liquid.
 ii Vapour irritant to eyes and respiratory system.
 iii Causes severe skin burns.
 iv Flammable.

2 **Planned Uses**
 i Very occasional use in dispensing for wart treatment.
 (**NB**: Very rarely used nowadays.)

3 **Unplanned Events**
 i Possible accidental spillage.

4 **Action Taken**
 i Product Safety Data Sheet obtained and made available to all staff. Appropriate instruction given in handling, storage, spillage, first aid, and other emergency procedures.
 or
 Dispensing staff made aware of risk information on product labels.
 First aid procedures:
 — wash eye with water for at least 10 minutes and seek medical attention.
 — drench skin with cold water, followed by the application of sodium bicarbonate. Seek medical attention where necessary.
 Spillage—absorb on inert material—eg Encap.
 ii Products stored in "flammables" cupboard.
 iii Appropriate fire-fighting equipment installed and regularly checked.

5 **Assessor:**

6 **Employer:**

7 **Date:**

FIGURE 11 *Sample Assessment, Glacial Acetic Acid*
(Source: National Pharmaceutical Association)

department must require suppliers to notify any changes in formulation, and there must be no deviations in purchasing from the lists approved in advance.

All staff should be informed of the assessments and invited to give their views about them. Necessary training in the principles or practice of their application should be given.

The small retailer must find ways of covering the same ground but of doing so with much more limited resources. It is necessary for him to rely to a greater degree upon his trade association for information and advice—and also upon his suppliers (particularly, for example, if they include a 'mutual' organization for buying in conjunction with other retailers, or a 'Cash and Carry' chain).

Most of the contributors to this volume have advised erring on the side of caution and preparing a written assessment in every instance, even though, strictly speaking, this is not required under the Regulations. While in retail business as well it may be possible for principals of large organizations to

Sample Assessment

The Control of Substances Hazardous to Health Regulations, 1988

INDUSTRIAL METHYLATED SPIRIT

1 Nature of Hazard

- i Flammable liquid.
- ii Irritant to skin and eyes.
- iii Vapour harmful if inhaled.

2 Planned Uses

- i Occasional dispensing.
- ii Supply to local surgery as:
  (a) original, unopened 2 litre bottle, or
  (b) Hibitane in 70% spirit.
- iii Occasional cleaning—eg oily residues from ointment slabs.

3 Unplanned Events

Possible accidental spillage.

4 Action Taken

- i Product Safety Data Sheet obtained and made available to all staff.
  or
  Dispensing staff made aware of risk information on product labels.
  Appropriate instruction given in handling, storage, spillage, first aid, and other emergency procedures.
- ii Product stored in locked "flammables" cupboard.
- iii Appropriate fire-fighting equipment installed and regularly checked.
- iv First aid procedures:
  - wash from skin with cold water.
  - wash eyes with water and seek medical attention.

5 Assessor:

6 Employer:

7 Date:

FIGURE 12 *Sample Assessment, Industrial Methylated Spirit*
(Source: National Pharmaceutical Association)

cover themselves by doing this, for many of the small people it would appear to be impracticable. An inspector could require the single-outlet retailer to be able to show documents (containing data in printed or hand-written form) on all the substances at his premises (for the most part these should be obtainable from suppliers) but it might be acceptable that the assessments were made to cover groups of substances—perhaps even, in some instances, that assessments were informal and not written down.

In this connection, it should be noted that besides the incoming goods, waste and scrap materials of all kinds should be assessed.

Attention should be given also to the training and to the attitude of mind of staff. It is relevant in this respect that when we enquired at a bookseller in the North of England, an agent for HMSO, mentioning the COSHH Regulations the assistant started to laugh (thinking, no doubt, 'more silly Regulations from London'). One point especially the retailer must ensure (not only when the inspector calls) is that all staff take the Regulations seriously.

Sample Assessment

The Control of Substances Hazardous to Health Regulations, 1988

FORMALDEHYDE

1 Nature of Hazard

i Vapour irritant to respiratory tract and eyelids.
ii Solution may cause skin sensitization.
iii Solution can cause severe eye-burns.
iv Explosive mixture can be formed with air in part-empty containers.
v Above 30% classified "Toxic" in CPL regulations.
From 5-30% classified "Irritant" in CPL regulations.

2 Planned Uses

i Occasional dispensing only (usually involving dilution. Possibility of spillage/skin contact etc).

3 Unplanned Events

i Possible accidental spillage.
ii Possible build-up of "explosive mixture" in unused stock containers.

4 Action Taken

i Product Safety Data Sheet obtained and made available to all dispensary staff.
or
Dispensary staff made aware of risk information on product labels.
Appropriate instruction given in handling, storage, spillage, first aid, and other emergency procedures.

5 Assessor:

6 Employer:

7 Date:

NB: Formaldehyde labels should carry the relevant information including the first aid information for eye-splashes, for example. If the manufacturer has not labelled it appropriately, you should obtain the manufacturer's Safety Data Sheet.

FIGURE 13 *Sample Assessment, Formaldehyde*
(Source: National Pharmaceutical Association)

Making a Critical Survey

Packaging is a much-criticized feature of modern retailing and no doubt in the future there will be pressure for less elaborate and sophisticated methods of sub-dividing and protecting the goods on display. Meanwhile, complex mixtures of packaging materials are a most important item on the shelves of most types of retailer; in general terms, while protecting the contents against adulteration or contamination, they present two kinds of hazard—physical and chemical.

Formerly, the grocer would arrange displays—making pyramids or other geometrical designs with filled cans or bottles of goods of various kinds. Now such arrangements are seen rarely; on the contrary, the packs are placed carefully in suitable shelving, often only one deep, with much reduced risk of collapsing and of bruising arms and legs, or causing cuts from broken glass.

However, while the physical problems associated with the pack are

Sample Assessment

The Control of Substances Hazardous to Health Regulations, 1988

SHARPS WASTE

1 Nature of Hazard

i Blood-contaminated needles and syringes from injecting addicts.
ii Lancets and other blood-contaminated equipment from cholesterol testing service.
(The hazard is mainly blood-born micro-organisms including HIV virus and hepatitis B virus.)

2 Planned Uses

i Addicts taking part in agreed scheme to exchange used syringes and needles for new clean ones.
ii Lancets used for producing blood samples for cholesterol testing service.

3 Unplanned Events

i Addicts arriving when pharmacists not available.
ii Contaminated syringes, needles and lancets being dropped.
iii Puncture of containers.

4 Action Taken

i NPA Guidelines and RPSGB* Guidelines obtained and made available to all staff participating in scheme (ie pharmacist proprietor, locums, managers etc).
ii Code of Practice devised and made available for ALL staff (to make clear that assistants should NOT participate, but giving clear instructions for what to do when pharmacist is unexpectedly absent).
iii Pharmacy disposal box stored in locked cupboard when not in use.
iv Heavy gloves, forceps and strong bleach available in case of spillage.
v Operating staff advised to have Hepatitis B vaccination.

5 Assessor:

6 Employer:

7 Date:

* Royal Pharmaceutical Society of Great Britain

FIGURE 14 *Sample Assessment, 'Sharps Waste'*
(Source: National Pharmaceutical Association)

reduced in ways such as this difficulties still could rise when items are removed from higher shelves, and in other situations (as at counters when purchases are transferred to baskets or boxes).

Packs with plastic or metal tabs which can be detached easily should be out of reach of children. Young people, inebriates, or 'simple' people also might harm themselves by choking on plastic film (perhaps even suffocate themselves, or another unfortunate person, by putting air-tight containers over their heads).

When packing is being disposed, all metals and glass should be separated. Incineration should be used to destroy plastics and coated papers only if suitable precautions are taken. The incinerators used should be as specified for the work, sited with care, and operated in accordance with specifications.

Reference has been made to the important changes which have taken place in the organization and methods of retailing. Notwithstanding the consolidation that has occurred, many specialist retailers remain and an important comparative novelty in this area in Britain has been the emergence of the 'mini-market'—a kind of modern general store in which a wide variety of goods of different kinds is stocked. (Mini-markets may be found on their own or attached to small garages, farm shops, or other similar sales outlets.) The goods can be very varied and often include foodstuffs (especially bread), household cleaners, paraffin, motor oils, ice cream, toys, and proprietary pharmaceuticals such as cough medicines and throat lozenges. Similar mixtures of goods on display can be seen on market stalls and (to a more limited extent) in the 'damaged' or 'special offer' sections of large multiples. There are obvious possibilities for contamination—arising from breakages, from defects in packing, from members of the public handling goods and putting them back in the wrong places, and so forth. Customers have been seen sometimes to 'sample' goods—by taking only part of a bunch or pack on display, and by putting sweets, fruit, and other foodstuffs in their mouths. Owners and supervisors of such markets, variety stalls, *etc*., must always be alert to such risks and ready to deal with them promptly.

Customers in retail establishments are discouraged (or, increasingly, forbidden) from smoking cigarettes and tobacco in other forms, but still these remain important possible sources of contamination in many outlets, including the 'covered market'. Smokers may leave behind spent matches, discarded dottle or ends, and a shopkeeper or stallholder has to be conscious of risks arising from such items—as also with many other pieces of detritus and artefacts—including small wheels or other parts lost from toy cars, used bandages, dolls' hair, and so on.

The general standard of cleanliness in food shops is fairly good (there are exceptions) but it is not always so elsewhere—as examples, in some charity shops, second-hand dealers, and second-hand bookshops. Sometimes premises are old and unsuitable, or maintenance is neglected because of costs. Vegetables and other green-groceries inevitably are accompanied by certain amount of earth, possibly stones, and some other items from the fields. However, under the Regulations, micro-organisms, dust, and dirt are all open to be considered 'substances hazardous to health'.

Most shopkeepers will be well aware of the dangers associated with 'foreign bodies' or 'taint' in the goods they sell, and will take reasonable steps to avoid them. However, to some extent they are in the hands of their manufacturers or wholesalers—for although it is perfectly clear, say, that a wedding ring in a can of peas must have been canned with the peas—a taint may have been acquired at any one of many stages. Even the 'foreign body' at times can be difficult to explain (if, for example, metal alleged to be in a sweet is identified as a filling from a tooth). As in all these matters, the retailer can only be on his guard.

Safe storage and handling of goods for retail sale is a topic in itself but matters to be reviewed in connection with these Regulations might include:

(i) general security
(ii) opportunities for tampering or sabotage

(iii) ambient conditions (the presence or absence of direct sunlight, whether natural light should be excluded, arrangements for heating and ventilation, the possibilities for malfunction or failure of control systems, the length of time in storage, the likely effects of long periods of heat or cold, and so on)
(iv) the suitability (or otherwise) of the conditions of storage and handling for the goods concerned
(v) risks of contamination (by other substances, micro-organisms, or mould)
(vi) risks of physical deterioration
(vii) risks of decomposition and of the decomposition products
(viii) risks of re-actions and inter-actions
(ix) risks of infestation
(x) the convenience and safety (or otherwise) of drawing goods from storage and transferring them to the point of sale.

When in each case a suitable assessment has been made, any corrective action that may be necessary can be taken.

In all aspects of his critical survey the retailer will experience difficulty—in obtaining relevant hazard data from suppliers, in assessing the risks associated with 'unknown' substances, in 'making-up' products for sale, and generally in foreseeing the future and anticipating unfortunate events. He can only go round his establishment with very clear eyes, and be always cautious.

Acknowledgement. The editors acknowledge the assistance of the National Pharmaceutical Association.

CHAPTER 10

Application in a Research Environment

J. P. SAMUELS

Introduction

The following section details the experience of BP's Sunbury Research Centre in applying the Regulations in a research environment. The author in no way lays claim to the systems developed and arrangements put into place as being his own. The success of the programme was achieved by a team effort from safety, medical, engineering and computing and information branches of BP Research, with the invaluable advice of the BP Group Occupational Health Centre. The importance of the enthusiasm and commitment of all levels of management and research staff in providing the necessary resources and actually carrying out the many hundreds of assessments should not be underestimated.

Research Activities within the BP Group

Many people associate BP with our traditional core business of oil exploration, drilling, and refining, as well as our interests in the chemical industries. In addition to these, however, we are engaged also in many other business areas such as the production of advanced materials, (polymers and reinforced metals), solar power, nutrition, and gas technology. It is necessary for all of these operations to be underpinned by a comprehensive research programme, which not only provides a service to the industries of today, but also looks to identify and to develop the industries and technologies of tomorrow.

The nature and scale of the work undertaken within this programme is incredibly varied. It ranges from small-scale laboratory operations following internationally standardized test methods, or the laboratory development of novel chemicals, materials and processes, to the bulk handling of hydrocarbon fuels for testing in engines, or the operation of scale-up pilot plants.

It is not difficult to appreciate that the range of hazardous substances handled, and the nature and extent of peoples' exposure to them, is very broad indeed.

The successful application of the Regulations on the site entailed two main problems. The first was related to the nature of the research activities themselves, with their own characteristics, which set them apart from other industrial sectors. The second was the sheer size of the site. With some 2300 employees, and being one of the largest industrial research establishments in Western Europe, there was a clear need to develop suitable systems to ensure that the programme could be managed on a continuing basis. These aspects of our implementation should be of interest to any large organization committed to the effective and efficient implementation of COSHH.

The Nature of Research

The nature of the research environment gives rise to some special problems in the application of health and safety legislation.

Firstly, as I mentioned earlier, there is the immense variety and diversity of the work. Many substances are used only once; many tasks are carried out only once. Other substances are handled far more intensely, with more continuous exposure, across the site. Any assessment procedure would need to be flexible enough to allow a 'suitable and sufficient' assessment to be carried out on all activities.

Secondly, the use of many novel substances, and the continually growing list of all chemicals used, meant that we would need to have a very flexible approach to providing health and safety information about them throughout the site.

Thirdly, until the relatively recent introduction of the Health and Safety at Work Act 1974 and its subordinate regulations, the research environment has not been accustomed to the direct application of national legislation. Few codes of practice relating to specific hazards have ventured to include research and development within their scope. Close ties with academic research carry with them as many problems as benefits in trying to instil new approaches to health and safety, and new ways of doing things.

Fourthly, the justification for the application of the Regulations on the site was not immediately obvious. We do not see research chemists suffering extensive mortality as a result of their occupational exposure. Epidemiological data do not, at first sight, appear to show any problems either. Let us look a little closer, however, and we will see that our inability to attribute specific causes to occupational disease should not lead us into complacency.

TABLE XIII *Risk ratios for various cancers, petrochemical industry*

Site of cancer	*Risk ratio*	*Confidence limit 95%*	*References*
Bladder	0.65	0.5 to 0.8	1, 2, 3
Central nervous system	1.17	1.0 to 1.3	1, 2, 3
Colon	0.91	0.8 to 1.0	1, 2, 3
Liver	0.92	0.7 to 1.1	1, 2, 3
Lung	0.87	0.8 to 0.9	1, 2, 3
Lymphoreticular	0.92	0.7 to 1.2	1, 2, 3
Rectum	0.85	0.7 to 1.0	1, 2, 3

Epidemiological Aspects of Occupational Cancer

Epidemiology is the study of disease patterns in groups of people.

Generally studies are carried out by comparing the frequency of disease in a group of people exposed to a suspected causal agent with that in a 'control group' not exposed to the agent. The figures then have to be corrected to allow for differences in the two groups, such as age, smoking patterns, socio-economic group, and dietary differences. The ratio of the first frequency to the second we can call the 'Risk Ratio', as it gives the chance of developing disease for a person exposed to the causal agent, in comparison with that for one who is not.

Before we look at specific problems in specific industries, let us first consider the scale of occupational cancer as a whole.

Approximately 150 000 people die in the UK every year as a result of cancer. Several attempts have been made to estimate the proportion of this total which is, in some way, caused by occupational exposure. Extimates vary between 2% and 30%, but even the most conservative studies indicate that the proportion is in the region of 4%. A mortality of 6000 per year arising from occupational cancer would put the problem on a par with that of deaths from road accidents. It is on this scale that it should be viewed, even without taking account of deaths attributable to other occupational causes, or of the pain and suffering associated with non-fatal diseases (morbidity).

So what of the Petrochemical Industry (Table XIII)? Many studies have been carried out, and these would appear to prove, on first sight, that there is no increased risk of occupational cancer in this industry. The Risk Ratios are all close to 1, and 1 is within the 95% confidence limits for almost all 'target organs'. It would appear to be a wonderful industry to work in if one wishes to avoid bladder cancer!

But what do we mean by 'Petrochemical Industry'? This very broad term will cover everybody in the industry from process operators and chemists in day-to-day contact with hazardous substances, to caterers, plumbers, carpenters and accountants, whose exposure will be considerably different. It could be the case that these figures mask problems associated with specific trades or specific substances, which are drowned amidst the 'noise' of the rest of the data.

TABLE XIV *Risk ratios for various cancers, chemists*

Site of cancer	*Risk ratio*	*Confidence limit 95%*	*References*
Bladder	1.26	0.8 to 1.8	1
Brain	2.4	1.8 to 3.1	1
Colorectal	1.45	1.3 to 1.6	1
Liver	1.31	0.8 to 1.9	1
Lung	0.63	0.6 to 0.7	1
Lymphoma	1.82	1.4 to 2.2	1

TABLE XV *Risk ratios for various cancers—exposure to 2-naphthylamine*

Site of cancer	*Risk ratio*	*Confidence limit 95%*	*References*
Bladder	2.68	0.3 to 10.3	4
Lung	1.03	0.5 to 1.8	4
Oesophagus	3.00	1.1 to 6.5	4

TABLE XVI *Risk ratios for various cancers—vinyl chloride*

Site of cancer	*Risk ratio*	*Confidence limit 95%*	*References*
Brain	1.37	0.9 to 2.0	1
Liver	4.68	3.3 to 6.5	1
Lung	1.12	1.0 to 1.2	1

If we turn our attention to chemists (Table XIV), we can see that there are slightly more grounds for concern. Almost all of the Risk Ratios have increased for the respective target organs. Perhaps more significantly, in many cases the lower bound of the 95% confidence limits is greater than 1.

Even in this case, 'chemists' will be employed in a wide variety of work from large-scale hands-on operations to administrative or management roles and in many different types of industries. In other words, there could still be a good deal of noise muffling and preventing the relevant data from emerging as strongly as it might.

Tables XV and XVI show significantly higher Risk Ratios for well-recognized carcinogens from various industrial and technical environments.

It is for the reasons described above that we cannot afford to be complacent about the scale of occupational disease in our industry. Our inability to attribute specific causes should not lead us to believe that they do not exist. We cannot know how many other strong cause/effect relationships lurk behind data which are available but come from very varied exposures.

COSHH Advisory Committee

A COSHH Advisory Committee was set up at the Research Centre in the summer of 1988 to advise on the implementation of the Regulation on the site. It was led by the site medical officer, and comprised other members from medical and safety groups, and the BP Group Occupational Health Centre. Three working parties were set up to advise the Committee on the development of an assessment pro forma, the setting up of computerized record-keeping systems and the training requirements for the site.

In the course of 1988 a pilot study was carried out within one of the research divisions to assess the operational and financial implications of the Regulations. By the end of the year we were in a position to present to all levels of management justification for our implementation plan. The general manager issued a site policy notice which identified compliance with the Regulations as a site objective for 1989. Money was allocated to enable the development of central systems, and a COSHH co-ordinator was nominated by senior management to undertake primary responsibility for the application of the Regulations.

Site Chemical Inventory

The duty to ensure that employees were provided with hazard data for substances was with us long before the introduction of COSHH but clearly, we were going to have to be much more vigorous in carrying this into effect. Initial investigations showed that around 25 000 chemical substances were in use on the site, so preparing a Site Inventory would have to be done very efficiently too.

Rather than allow individual operating groups to pursue suppliers independently, it was decided that the best solution would be to set up a central data base of all relevant substances and their health and safety information. With this in view, a health and safety information officer was appointed.

Several commercially-available computer systems were considered, but for our purposes all had two main drawbacks: firstly, they did not have the facility to write new substances and new information into the system; secondly, they did not allow the information to be loaded on to a mainframe system and to give access to others elsewhere on the site. In response to these problems, we decided that there was no other solution but to compile our own data base.

With the help of our on-site computer consultants the RICH (*R*esearch Centre *I*nventory of *C*hemical *H*azards) data base was established.

The system records the name and reference number of each substance used, the groups which use it and their locations, broad user-type categories, and health and safety data relating to the substance. Apart from the obvious benefits of the system for our COSHH programme, there are other operational benefits. For example, a group which requires a small quantity of a substance can locate other users on the site. This may make it unnecessary for those concerned to order further quantities of the substance, much of which could well remain unused so that we would have to pay to dispose of it in future.

It is anticipated that the data for all substances will be entered in full by the middle of 1991.

With a view to maintaining the data in the future, procedures have been built into systems for purchasing so that any new substances coming on to the site for the first time are identified and entered.

Generic Assessments

Perhaps the most difficult, and the most interesting, aspect of applying COSHH in a research environment is the optimum division of work into assessable units. The very nature of the research work means that it is simply impossible to carry out a separate assessment on every task that is to be performed with every different substance. Also, this would be counterproductive if at the end of the exercise the information is to be read, understood, and be useful to employees. The assessment pro forma we have developed is as relevant and useful in the catering, workshop, and office environments as it is in the laboratories, but the key to using the pro forma efficiently is flexibility. No one approach is right for all environments: the approach must be modified to suit the environment.

Let us imagine that we are given the job of painting a hundred pieces of garden equipment—ten spades, ten forks, ten picks, ten hoes, ten rakes, and so on. Whilst the paint for all of the items has the same solvent base, the matter is complicated further by the fact that one of each type of equipment is to be painted a different colour. How then would we go about carrying out a COSHH assessment prior to starting the work?

Certainly the nature of the work is no different for each of the items. They are all of similar size, have similar surface areas and would need a similar amount of time to be painted. From the point of view of the protection of health the different colours of the paints present no difficulties either. Whether we were using red, blue, or yellow paint we would probably specify that the work should be done outside, with precautions being taken to prevent skin contact. There is no need to carry out a a hundred different assessments along the lines of, 'paint spade red, paint spade blue . . ., *etc.*'. We can identify a single assessment called 'Painting garden implements with solvent-based paints'. We will need only to identify further assessments if the work changes such as to require different control measures;—as examples, should we wish to treat with a pest-controlling substance, or to paint the same implements indoors.

Whilst this may seem to be a trivial example, the approach to assessing real work activities should be very much the same, particularly in the research environment. If we wish to carry out ten different operations with ten different chemicals, and the control measures required will be the same in each case, then we need only carry out one assessment. It is only when one of the parameters changes, such that there is a change in the control measures necessary, that we need to carry out further assessments.

Early estimates indicated that across the site approximately 1500 assessments would be required. However, by making use of GENERIC ASSESSMENTS, this was reduced to around 650 actually having to be carried out.

Responsibility For Assessments

A question that every employer has had to answer is 'Who is competent to carry out assessments?' Some have set up assessing groups comprising key personnel to carry out all of them. Others have placed the task in the hands of occupational hygienists, either their own employees or from an external consultant.

We decided at a very early stage that responsibility for carrying out assessments should be with line managers. Certainly they would require training, and there would need to be some sort of checking and auditing by site specialists, but there are several fundamental reasons why we thought this approach to be the most appropriate.

Firstly, the fact that assessments are carried out locally means that staff are involved immediately in the process of gathering and assessing the information. This means that at a later stage it is much easier to ensure that they receive and absorb the conclusions of the work.

Secondly, operating groups are far more likely to comply with rules and to fulfil actions that they have drawn up themselves, rather than having them imposed by a central group.

Thirdly, it would be impossible to manage the system, and the continuing need for assessments, unless operating groups had the COSHH requirements integrated fully with their other operational priorities.

The assessments completed by line managers were of a remarkably high quality, and we would have no hesitation in using the same approach again.

Carrying Out Assessments

The first task we set for line management was to divide the work in their areas into assessable units. Staff concerned were encouraged to use the generic approach, and asked to list the titles of assessments on a COSHH 1 Pro Forma. (This serves as a developing index to the assessments.)

Many managers asked for advice and help to achieve the optimum division of work into assessable activities. Ultimately most groups found that they had less than eight assessments to make.

Once the necessary assessments had been identified, managers concerned were asked to carry them out using the COSHH 2 Pro Forma which had been developed by the site specialists. They were encouraged to come whenever necessary to safety and medical groups for advice. Once the pro forma had been completed it was required to be signed by an appropriate level of management in the operating group concerned in order to approve the actions which had been allocated. Then it was required to be checked for consistency by a site specialist to ensure that the conclusions reached and actions allocated were commensurate with the information recorded about the activity. After that had been done, the actions were recorded on the management system, which would help us to check the actions at the appropriate time.

The Assessment Pro Forma

The requirement of Regulation 6 of COSHH is for the assessment to be 'suitable and sufficient'. The Approved Code of Practice goes on to explain that an assessment can be considered suitable and sufficient 'if the detail and expertise with which it is carried out are commensurate with the nature and degree of risk arising from the work, as well as the complexity and variability of the process'. We must have this very much in the forefront of our minds if we are to produce assessments of the necessary quality, without committing unnecessary time and effort. The use of an assessment pro forma certainly helps in ensuring that this task is carried out efficiently and effectively. It acts as a check-list, and provides a degree of consistency between operations and work groups.

The pro forma which has been developed is virtually the same as that used in the rest of the BP Group, and probably similar to those being used by other large employers who have considered this approach worthwhile.

The process of assessment itself is the same for any activity. The pro forma encourages a step-by-step approach along the following lines:

(*a*) Divide the activity into discrete tasks which are to be assessed separately. This does not require such a detailed division as would be required for a Task Analysis; tasks should be divided only if the substances, inventories, or operations differ to an extent such as to require different control measures. Hence, if a container of 'Solvent A' is to be opened, and the contents decanted and weighed in a fume cupboard with no change in the control measures, then this should be considered as one task—'Handling of small quantities of Solvent A'—and not three;

(*b*) Give a brief description of each task, and record basic information about it which may be relevant to the assessment; as examples, the number of people involved, the location, frequency, and duration of the task;

(*c*) List the substances to be used or present, with basic information about them which may be relevant to the assessment; as examples—in what physical form will they be, how much will be present, what health and safety data sheet references are applicable, are there any occupational exposure limits for the substances, what are their main effects on health, can they be absorbed through the intact skin?;

(*d*) List the control measures which are to be employed. Where there are *procedural* controls, then any reference numbers of standard methods, operating procedures, permit-to-work systems, *etc.* should be recorded. Where there are containment, ventilation, or other *engineering* controls, then their reference numbers also should be recorded. Reference to any item of 'PPE' (Personal Protective Equipment) should include the site approval number, so that its suitability can be checked;

(*e*) Assess the possibility for exposure during normal operations for each substance, with regard to each route of entry (inhalation, skin contact, ingestion). We have found it useful to assess this possibility for exposure ignoring the protection afforded by any PPE (this serves to emphasize the extent to which any PPE is being relied upon, and whether other control measures should be provided instead). Often it is sufficient for this assessment to be made on a qualitative basis, with no need for measurements of airborne concentrations. We would simply assess the possibility of exposure as being 'high', 'medium', or 'low', using the following criteria:
HIGH—Exposure close to or in excess of any occupational exposure limits, or unnecessarily high, likely to occur on occasions;
MEDIUM—Some exposure likely. Well below any exposure limits; and
LOW—Very little exposure.
Any data and other supporting information on which these assessments have been based should be recorded; as examples—observations and experience, medical reports or symptoms, hygiene, and other reports;

(*f*) Record details of any facilities for response in event of emergency. These should include references to any written procedures, and to any equipment provided for use in an emergency;

(*g*) Record any other hazards observed that may require further investigation. Although these may not be covered by the COSHH Regulations, other problems such as ergonomics or ionizing radiations may have been noted whilst carrying out the assessment. Clearly it is sensible to record them so that they can be followed up;

(*h*) Assess the risks associated with each task. Our pro forma allows one of three conclusions to be drawn:
RISKS INSIGNIFICANT—No control measures required;
RISKS NOT SIGNIFICANT BECAUSE OF EFFECTIVE CONTROL MEASURES—No additional control measures required, but continuing action needed to ensure control measures remain effective; or
RISKS SIGNIFICANT—Additional control measures required or current control measures inadequate.
Once this assessment has been made, then the actions required can be determined; and

(*i*) Record any actions required. Allocate *personal responsibility* for ensuring that each action is completed, with target dates where appropriate. These actions can be identified by working through regulations 7 to 12, plus any other hazards that may have been noted while making the assessment.

With the help of a well-designed and completed pro forma the reader can build up a good mental image of the activity, and identify any errors or inconsistencies. Any modifications to the review must be agreed with the assessors and managers before being signed off centrally and entered on to the administrative data base.

Administration of the Assessment Pro Formas

On a site of size large enough to generate several thousands of initial actions, and probably several hundreds on a continuing basis, it clearly is important to have an effective system of administration. At the time of setting up our systems for COSHH we were blessed with the presence of a site medical officer with both a specialization in dermatology and a sense of humour. I understand he spent many sleepless nights thinking of an appropriate acronym, which any self-respecting computer system, of course, must have. Ultimately, the RASH (*R*esearch *A*dministration *S*ystem for COSH*H*) system was born.

The RASH system is designed to record actions arising from assessments, along with the target date for completion of the last action, and the person (or appointment) responsible for ensuring that the action is carried out. It does not attempt to record the details of assessments, although we may develop a computer-based package for the electronic recording of assessments in the future.

Operating groups are required to notify the RASH administrator on completion of the last outstanding action on each assessment. If the target date passes without notification being received a reminder letter requesting formal notification is generated automatically.

Control Measures

From the assessment pro forma we must establish three important facts about the control measures.

Firstly, we must establish that the control measures proposed are appropriate for the activity being carried out; that is, that they are capable of controlling the risks arising from the activity, and that they are pitched at an appropriate level in the hierarchy of control as described under paragraph 7 of the Approved Code of Practice. It may be that too much reliance is being placed upon personal protective equipment, and greater containment or local exhaust ventilation is possible, or that one of the substances involved could be replaced by a material inherently safer.

Secondly, we must establish whether the control measures are likely to be used in the manner described by the assessment, or are so cumbersome and inconvenient that they are not likely to be used. If they are to be maintained effectively, most procedural controls will need to be documented properly, either locally or by reference to a standard test method. The assessment should reflect the line management's recognition of the duty to provide effective supervision.

Thirdly, unless control measures are maintained properly, examined and tested on a regular basis it is pointless merely ensuring that they are used. The responsibility for maintaining procedural controls clearly lies with line management, to ensure that they are reviewed periodically, and in particular when significant changes take place in the work. The arrangements for engineering controls and PPE are covered in the two sections following.

Engineering Controls

When we set out on the road to establishing systems to comply with the requirements of the Regulations, it was clear that one of the most difficult tasks would be to establish and to maintain records of the regular examination and testing of engineering controls. To illustrate this, we have on the site approximately a thousand ventilation systems, as well as many systems for detection, alarm, and containment. Also, the COSHH requirements were not arriving in isolation. Regulations concerning pressure systems and electricity at work were bringing the direct application of many areas of national legislation to the site, and bringing with them additional requirements for the regular examination and testing of equipment (besides the established checks of items such as lifting equipment and certain pressure vessels). The requirements for record-keeping associated with all this were daunting.

Our response to the problem was to set up a site inspection group, responsible for ensuring that regular examinations and tests of equipment were carried out, and suitable records maintained. All fume cupboards and other ventilation systems were tagged with identity numbers, and base-line tests carried out (in some cases indicating the need for remedial work). These regular tests demonstrated a need to review procedures for specifying performance standards of new ventilation systems, providing criteria against which they can be tested in the future. Where the Approved Code of Practice does not indicate frequencies of examination for specific types of equipment, we have had to decide ourselves upon 'suitable intervals' with the help of international, national, industry, and company codes of practice. Currently, steps are being taken to enter maintenance records on the same system.

Personal Protective Equipment

Among other things, implementing COSHH required the introduction of a comprehensive programme for PPE.

Approved lists were drawn up for the various types of equipment in use, and we are preparing information sheets giving the details and extent of the approvals. Items held on stores stock have been reviewed, in many cases after trials by users, with a view to increasing the number of operations covered and numbers of staff who can be fitted as standard. When complete the information sheets will be issued with the equipment, to ensure that they reach points of use.

In the case of 'RPE' (Respiratory Protective Equipment) training courses have been organized for users to ensure their competence to fit, use, clean, maintain, and check the equipment on a regular basis. For positive-pressure RPE, a system of monthly testing by a competent person has been established.

Routine Monitoring

In some cases, monitoring of the atmosphere at the place of work may be appropriate. In the first instance it may be necessary in order to carry out a suitable and sufficient assessment of the risks to health arising out of the work activity, under Regulation 6. In the second instance, monitoring may be necessary on a regular basis to detect any deterioration in the control measures which could lead to a serious effect on health, or where it is necessary to ensure that occupational exposure limits are not exceeded. Monitoring on a regular basis is a statutory requirement for vinyl chloride monomer and for many electrolytic chromium processes.

We have not found the need for an extensive programme of routine monitoring. In fact, monitoring has been very limited. It seems preferable to employ the limited resources of time, money, and effort on matters such as making thorough assessments, appropriate control measures, sound maintenance, testing and operating procedures, and comprehensive training programmes. It is felt these should be the primary methods of approach. If you find within your activities an extensive need for routine monitoring, I would suggest that you go back and look again at these aspects of control. Routine monitoring is no substitute.

Some examples of activities in which we have found it appropriate to carry out routine monitoring are those requiring extensive use of mercury and certain aspects of semi-conductor work.

Currently, records of the results of routine monitoring are being maintained in traditional form: we plan in future to develop electronic record keeping.

Health Surveillance

Much of what has been written already about routine monitoring applies equally to health surveillance. It is very much a secondary safeguard, and is no substitute for proper control measures.

Regulation 11 requires that health surveillance be carried out where there is a 'reasonable likelihood' that an identifiable disease or effect on health will occur, and where there are valid techniques for its detection. Such situations should be few and far between.

Where health surveillance is employed, its nature will vary: examples include: the maintenance of health records for some work with carcinogens, skin inspections for exposure to used engine oils, and regular biological monitoring for employees who work extensively with mercury or other heavy metals.

Training

Effective training is the most critical aspect of any implementation programme. It would be pointless to introduce new systems and methods, without teaching employees how to use them properly. (Unfortunately, sometimes it is left until the programme has started to collapse, and everyone concerned has become thoroughly cynical, before a rescue mission is attempted.)

The principal commitment has been to the training of line management. The introduction of the Regulations presented an ideal opportunity to educate managers and operating staff in matters of occupational health. A series of one-day training courses was held, covering the principles of occupational hygiene and basic toxicology, together with the historical, legal, and epidemiological background to the Regulations and requirements. Half of the day was devoted to case studies in assessment. The need for appropriate training may well vary considerably between, say, an office-based area, an engineering workshop, and a research laboratory. With this in mind, rather than attempt to dictate from the centre the numbers or levels of staff who should attend this course, the onus was put upon line management to ensure that there were enough trained personnel within their areas to enable them to carry out initial assessments and to maintain the system in satisfactory condition.

Over 350 staff attended the initial phase of training courses, far more in number (and motivated far more positively), than if they had been summoned.

Other training carried out under our COSHH programme included briefings for safety representatives, for users of fume cupboards and PPE, and *ad hoc* presentations for staff generally. The site arrangements for complying with the Regulations have been incorporated into induction training courses for all staff, and in particular for newly-appointed team leaders. The specific on-the-job training for staff has been reinforced and formalized, but remains very much a line management responsibility.

User Information

The generation of information and records for COSHH is not an end in itself but a means to improve the control of substances hazardous to health which are used in work activities. However, this will happen only if the information generated actually reaches the people who use the substances, and in a form which is convenient and they can understand.

To this end, all work groups are issued with a COSHH File—in which are kept health and safety data sheets, COSHH assessments, HSE and other guidance notes, and various information related to occupational health relevant to the group's activities. Within the work areas, such files have proved to be very useful focal points.

Publicity and Promotion

I have stressed already the importance of management commitment to the success of our COSHH Programme, but the enthusiasm of operating staff who actually use and may be exposed to the substances concerned is at least as important.

Publicity campaigns have taken a number of forms, at various stages during the implementation plan. They have included the showing of videos, posters, and sending information by post.

Auditing

Once the initial phase of assessment has been completed and the system is up and running some form of auditing is necessary to monitor performance. We have started a monthly audit system in which we select a work group at random, and look through all aspects of their storage, handling, use, and disposal of hazardous substances, as well as their health and safety data sheets, COSHH assessments, operating procedures, and other documentation. Broad conclusions of each audit are given a wide distribution, after consultation with the relevant line manager, so that the lessons learned can be applied across the site. (In the early days of these audits we learned from them as much about the adequacy of our own systems as we did about the efforts of the group we were auditing.)

The Health and Safety Programme

A commentary on the implementation of the Regulations at Sunbury Research Centre would not be complete without reference to the second major implementation plan which we undertook concurrently with COSHH. This was the introduction, along with the rest of the BP Group, of the ISRS (International Safety Rating System) health and safety programme and auditing system.

Briefly stated, the system involves the setting of measureable objectives within a framework of twenty programme elements, formulation of a plan to achieve those objectives, implementation of the plan, and monitoring against the initial objectives.

The results of the programme have been outstanding. It has set up an infrastructure of group meetings, management inspections, training, regular promotional campaigns, and organizational rules which encourage a structured and co-ordinated approach to the management of occupational health and safety. It has set performance standards for all levels of supervisors and managers, incorporated health and safety performance into all job descriptions and staff appraisals, and brought about visible commitment to health and safety from the highest levels of management. Most importantly, it has harnessed the energy and initiative of the members of the health and safety team all-too-often forgotten—the line managers and operating staff who bear the ultimate responsibilities. It is in the winning of the hearts and minds of these people that the most dramatic and lasting improvements in health and safety are to be found. With this infrastructure behind us, we can be confident of the continuing success of our COSHH systems in the future.

Conclusion

The resources we have committed to our programme for implementing COSHH have been considerable. We believe we have found the right flexible approach to make the process as efficient as possible.

Assessments have identified already many activities where control measures were less than adequate, and steps have been taken to put better control measures in place.

Examinations of control measures have indicated deficiencies which have been put right.

All these approaches have been built into our procedures on site and will become part and parcel of the way we approach our work in the future. There will be a continuing cost of operating and maintaining the procedures, but we believe this to be outweighed in terms of increased protection of the health and safety of our employees.

Acknowledgement. The author is grateful to the owners of copyright in the references shown below for permission to reproduce information.

References

1. M. R. Alderson, 'Occupational Cancer', Butterworths, London, 1986.
2. N. M. Harris, L. G. Shallenberger, B. S. Donaleski and E. A. Sales, 'A Retrospective Mortality Study of Workers in Three Major U.S. Refineries and Chemical Plants', *J. Occup. Med.*, 1985, **I.27**, pp 283–292; **II.27**, pp 361–369.
3. O. Wong, R. W. Morgan, W. J. Bailey *et al.*, 'An Epidemiological Study of Petroleum Refinery Employees', *Br. J. Ind. Med.*, 1986, **43**, pp 6–17.
4. F. B. Stern, L. I. Murthy, J. J. Beaumont, P. A. Schulte, W. E. Halperin, 'Notification and Risk Assessment for Bladder Cancer of a Cohort Exposed to Aromatic Amines'. *J. Occup. Med.*, 1985, **27**, pp 495–500.

CHAPTER 11

COSHH and the Air, Land, and Water Environments: An Integrated Approach

R. W. HAZELL, S. G. LUXON, and M. L. RICHARDSON

Introduction

The Regulations focus attention on a need which has been perceived for some time—the development of a better understanding on the part of the public of methods by which risks may be assessed and expressed in quantitative terms.*[,1]

On the other hand, assessors of risk need to appreciate that the acceptance of their work by the public is influenced by other factors which are not necessarily quantifiable. Bearing in mind some erroneous and misleading pronouncements in the past on the part of officials and other spokesmen there may be quite understandable scepticism. By now, the general public (including one's employees) is aware that as a result of increased knowledge and experience some substances that once were regarded as 'safe' actually require much more care—in disposal, handling, and use—both from the point of view of the safety of individuals and also in more general environmental terms, such as when being disposed as waste or as products of decomposition (such as, with large emissions of carbon dioxide or indeed, in at least one instance, of water vapour). Notable examples of changes in attitudes towards substances over the years include benzene, asbestos, aluminium, and vinyl chloride, but there are numerous others less well-known. Until quite recently, many of us probably were satisfied with a less

*Under EEC requirements the definition of risk shall always include a measure of quantification.

sophisticated view than is required now of the need to assess risk in terms that would be as objective as possible.

This is not to say that a more rigorous scientific approach can eliminate the influence of the unknown factor. Science is accompanied by uncertainty as well as the opposite. At the time of writing, attention has been drawn to a new study of possible effects on human reproduction of low levels of radiation (there may be unanticipated hazards).[2] Some effects reported might be shown to be spurious or from other causes but if this work is verified it could alter radically current views of what is acceptable in the nuclear industry, and elsewhere where radio-active materials are transported or used.

A specialist in risk still may need to convince others of the value of the techniques he uses—while at the same time recognizing that the foundation might not be bed-rock.

Essentials of Risk Assessment

For specific substances, risk assessment and management procedures can be applied in two different modes:

(i) individual hazards

(ii) combined hazards (as examples: chimney effluvia, engine exhausts).

To take each in turn—individual hazards generally can be dealt with more easily as data can be obtained from monitoring and other sources as to the quantities involved and effects anticipated at given concentrations. In such cases, risk management may resolve itself into reducing or eliminating entirely the use of a particular substance, perhaps by employing another less hazardous in its place.

On the other hand, combined hazards can arise from many substances in uncertain mixtures, and from the point of view of management there may be difficulties even in identifying every substance involved, let alone assessing their effects together.

At this point it may be helpful to clarify the difference between the 'assessment' and 'management' aspects. The former includes the correlation and interpretation of information, together with prediction as a basis for action to control and prevent hazard. More specifically, it is the summation of (using the approved terms when indicated by single quotes):

(*a*) the identification of hazard and its assessment

(*b*) 'risk characterization'

(*c*) 'assessment of exposure'

(*d*) 'risk estimation'

—while management encompasses 'risk evaluation', the control of exposure and monitoring—the practical actions taken to control or prevent. Figure 15 expresses these ideas diagrammatically.

Increasingly, major schemes for development of new industrial sites and existing facilities are being required to include so-called 'environmental impact assessments'—which, essentially, are assessments of the likely effects

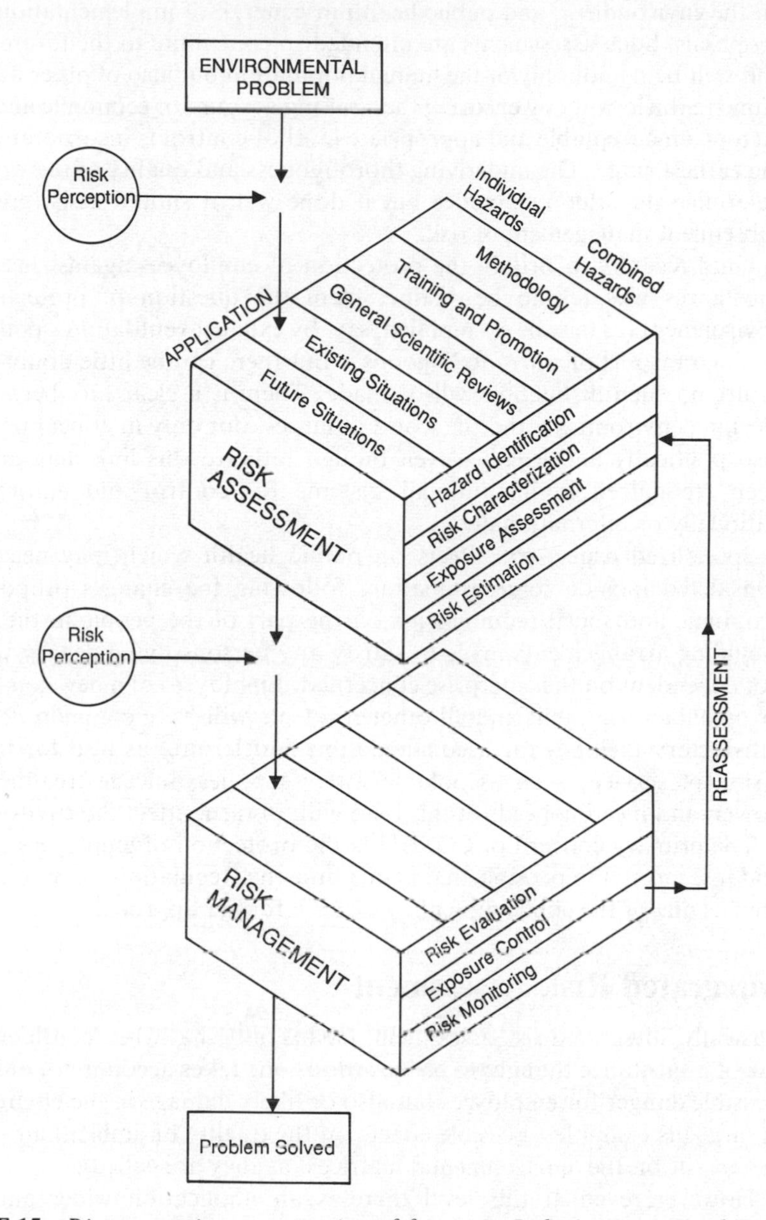

FIGURE 15 *Diagrammatic representation of factors in Risk Assessment and Risk Management*.
('Control of Environmental Hazards—Assessments and Management of Environmental Health Hazards', WHO/PEP/89.6. Reproduced with permission of the World Health Organization, Geneva)

on the environment, and public health in general, of implementation of the proposals. Such assessments are intended to contribute to the future health and well-being not only of the human population but also of other flora and fauna in the locality by ensuring that, taking account of economic and social factors, an acceptable and appropriate level of control is incorporated from the earliest stage. The underlying thoroughness and quality of the work will determine its value in practice but if done well it should assist greatly in subsequent management of risk.

Until recently in Britain the protection of employees against hazardous substances was felt to be distinct from the question of protecting the environment (as may be exemplified, say, by exhaust ventilation, spoil heaps, or the transport of slurry to 'lagoons'), but there can be little doubt that in future no such distinction will be made. There is a clear link between the working environment and the world at large—not only in ethical terms but also politically and legally—even though hitherto this link may not have been recognized in institutional systems for control and enforcement, nationally or internationally.

Specialized aspects of effects on public health which may need to be considered include some estimation, following the changes proposed, of economic and social requirements on the part of the people in the area—including arrangements made to satisfy any persons living nearby who are not dependent on the enterprise concerned. Employees of a new commercial or manufacturing unit and all other residents will have common needs for satisfactory facilities for accommodation and leisure, as also for the provision of services such as potable water supplies, sewage treatment and power, and meeting needs of this kind will, in turn, affect the environment.

The primary concern of COSHH is the protection of employees at work and it seems not impossible that in due time the Regulations may come to be seen as one of the concluding phases of the former approach.

Integrated Risk Assessment

Basically 'integrated risk assessment' means only that when considering the use of a substance thought to be hazardous one takes account not only of its possible danger for employees but also of likely damage in the environment at large (as examples, possible effects on the quality of ambient air, soil, or water—or on the 'environmental matrices' as they are called).

However, even at this level there is an implication wider and more profound than is being put forward in Britain at the time of writing in the Environmental Protection Act 1990.[3] The Act requires integrated control of pollution in all sectors of the external environment, and of course underlying this would be integrated risk assessment with a view to minimizing adverse effects, as for example in Case Study No. 3, pages 156–157. In this connection though it is not impossible to imagine an extreme case in which the needs of external environment and factory were conflicting: integrated assessment might suggest use within the factory of a substance of a relatively hazardous character on the grounds that it was more acceptable in terms of the external environment than a less-harmful substitute. This would be an uncomfortable conclusion, and—more to the point—would not be compatible with the Regulations.[4,5] If action along the lines proposed were taken the employees would have to be protected, possibly at considerable extra cost—and this would be done with a view to achieving, what, sometimes at any rate, may be less precise and obvious benefits.

Fortunately, although integrated risk assessment can raise fundamental problems such as this, in practice the choice usually will be less exacting. Simple 'good house-keeping' may well prove the most effective way of protecting both employees and the world beyond.

It will be helpful at this stage to state briefly the principal ways in which substances used at work enter the wider environment—that is, through transport and 'disposal' (in the broadest sense), to air, land, and water. Later, some consideration will be given to relationships between these routes and the use of chemical substances, in light of the concept of integrated risk assessment.

(i) Air Quality

The COSHH Regulations might be considered a first step towards limiting the pollution of ambient air by control of contaminants at the place of work, whereas the Environmental Protection Act extends the scope to the external environment. In this connection, the sections of COSHH dealing with maintenance of control equipment (such as fume cupboards) are of particular relevance. It may be necessary to provide treatment of the exhaust from such equipment (*e.g.* by filtration) and ensuring that stacks are high enough also may be important. In cases where very hazardous materials are being used there may be a need for more rigorous control (including, say, direct incineration of emissions in a suitable furnace).

(ii) Disposal of Waste

Waste material may be in a variety of physical forms (typically—solids, liquids, slurries, gases, or vapours—or mixtures of these) and may be grouped conveniently as:

(*a*) substantially a single product or by-product
(*b*) a random mixture of a small number of different substances
(*c*) complex mixtures or compounds.

The different methods of disposal include incineration under approved conditions, or burial at licensed sites. Frequently, disposal involves transporting the waste (by motor vehicle, railway, or ship) and in Britain work of this nature may be done only by licensed contractors. The consignments are required to be documented but there can be problems with loads comprising mixed wastes—when it may be difficult to establish a 'typical' analysis (mixed waste, of course, is not homogeneous) and so to identify the material, and where the composition in different drums of the same consignment may be quite different.

The analysis, testing, and assessment of the hazards of waste require further investigation and elaboration. Toxicological assessment on the basis of 'structural activity relationships' or 'structural analogy relationships' requires knowledge in some detail of the chemical nature of the substances in the waste. Tests of 'genotoxicity' have been found to give too many false results, both positive and negative. Other than in expensive tests *in vivo* there would appear only to be a tenuous correlation between mutagenicity and carcinogenicity. The approach via 'structural activity relationships' has been unreliable so far but it may be that in future a test based on induction P-450 cytochrome activity and steric factors would be worthy of investigation.[6]

The legal position currently is not quite clear but an approach being adopted increasingly is that unless shown otherwise any waste should be regarded as hazardous. In this scenario, the onus of proof rests with those consigning the waste; in the absence of evidence as to its safety, all material sent for disposal is regarded as 'special'.

(iii) Incineration

Currently, efficient incineration is the most satisfactory method of disposing of many forms of waste: unfortunately, it is also rather costly.

The incinerators used must be both designed correctly for the purpose and operated in accordance with instructions. Combustion temperatures must be high enough for safety (generally, of the order of 1200 °C), residence times must be adequate, and the stacks of sufficient height. Management procedures are required to ensure that approved practices are maintained, and to reassure the public at large. The possibility of washings from roofs and stacks finding their way into surface water and streams should be considered and if necessary action taken to prevent this.

Waste in some forms (*e.g.* domestic waste) has a calorific value high enough to justify burning it to generate electric power, and it may be that the ending of the monopoly in Britain in generating electricity will stimulate interest in this method of disposal. Another field in which technical development could be helpful would be through the introduction of smaller and more accessible incinerators. The disposal by burning of small quantities of substances such as waste solvents from laboratories could be acceptable, but it requires careful preparation in advance and adequate precautions.[7]

(iv) Landfill

The European Community has required for some years the completion of a programme of testing before permitting any new substance to be brought to market. This means that there should be data for such substances, possibly available in more systematic form than for some materials which have featured in commerce for some time. However, it is unlikely that the information will be of sufficient value in detail to permit assessment in the real environment.[8,9] In many instances, substances in contact with soil undergo anaerobic biodegradation, to form less harmful products of decomposition, but this does not happen every time. Many instances have been reported of landfill sites (sometimes many years after disposal has ended) where noxious gases were emitted when land was in use as car parks, or for warehousing or housing.

A proper scientific appreciation of the chemical and biological processes that are involved in landfill would require an understanding of anaerobic biodegradation somewhat greater than is available at present. At the time of writing it probably would be quite a generous estimate to say that one per cent of the substances listed in the 'European Inventory of Existing Chemical Substances' ('EINECS') had been investigated; only a few of these studies were undertaken under Good Laboratory Practice, and fewer still indicate the conditions under which degradation took place or the mechanisms of the reaction involved.

A well-planned and properly constructed site for landfill should be lined throughout with impervious material, to prevent the leaching of substances from the waste into water in the sub-strata. Failure to do this could result, in due course, in contamination of water supplies used for either irrigation or for potable supplies—with accompanying hazard to crops and health.[10,11]

The technique of encapsulation is considered a superior method of containing the waste, although it is expensive and leaching still can take place. Encapsulation and deep burial may offer advantages in the short term but in reality they only postpone the time at which the problem must be dealt with finally.

(v) Discharge to Water

Often, direct discharge to the aquatic environment is the most convenient and cheapest method of disposing of wastes. It may be done by long pipelines direct to sea, to rivers, to public sewage treatment works, or to an employer's own effluent facility—with subsequent disposal of treated material by one of the three routes mentioned earlier. Water used industrially for cooling, and containing low levels of substances not likely to be harmful, may be discharged to suitable soak-aways.

Both the carboniferous and nitrification processes in sewage treatment can be affected by chemicals in trade effluents and equally the chemicals affected by such treatment. However, the usual outcome for a substance entering a treatment works would be one of the following:

(*a*) the substance remained unaffected by the treatment [the usual words are 'persistent' (OECD) or 'recalcitrant' (Department of Environment)]
(*b*) there was partial biodegradation or transformation
(*c*) ultimate biodegradation—to water, carbon dioxide, sulphates, halides, nitrates, *etc*.

Probably the most difficult situation was that of partial transformation, since the metabolites might be unknown; in some instances, relatively innocuous chemicals could well be transformed into others more hazardous.

Many chemicals can be removed by absorption into sludge, which then would be subjected to anaerobic digestion before (say) being spread on agricultural land or discharged at sea. Unfortunately, some hazardous chemicals interfere with anaerobic digestion.

Discharge to rivers, whether directly or after treatment, can affect adversely invertebrates and fish. In testing for toxicity and ecotoxicity this is important, since usually differences between species in metabolism and the mechanisms by which damage occurs are quite significant. As an instance of this, some substances not regarded as hazardous for rodents or man can be acutely toxic to fish and to invertebrates (and the converse also is true).

Many water supplies involve the re-use of water from rivers—again perhaps for irrigation or for drinking. The differences between effects of the same substance on different species can be important in the sense that some substances thought unlikely to be harmful to man or to other mammals are deleterious to plant life.

In less-developed countries, even when treatment plants are available at a factory, their use may be restricted through problems with maintenance or the cost of power—and with units making fertilizer such difficulties can lead to prolific growth of vegetation in unexpected places.[12]

First Steps Towards Integrated Risk Assessment

In the early stages the most likely outcome of the integrated approach would be little more than good housekeeping—that is to say, the establishment of procedures to protect employees and to keep to a minimum the generation of waste and impairment of air quality.

In general terms, the disposal of waste is perceived as costly and the wastes themselves as commodities with negative value (in some instances, extremely expensive liabilities). Ultimately, in the overall view, it might be advantageous to increase the costs of control within the unit and/or to recycle waste so far as possible rather than to seek to dispose of it. Especially this would be so if the means of disposal were to an aquatic environment in which the substance or substances concerned, and perhaps harmful products of degradation, might accumulate and multiply. In this connection, it is anticipated that the European Commission will in the near future institute labelling requirements for 'Substances Dangerous to the Environment'.[9]

It is easy to advocate the enhancement of measures for protection but not always easy to justify it. There may, for example, be resistance from the operators involved. In the last analysis the acceptance of integrated risk assessment and its application must be common ground for the employers, politicians, regulators, and the public in general—and indeed unless such an approach is adopted, with appointment of the necessary 'integrated' inspectorate, prevailing conditions are not likely to be much improved.

Until that time, there are a few preliminary measures that could be taken.

As an instance, the 'Authorized and Approved List' as one of the legal requirements in Britain for the classification, packing, and labelling of substances provides risk and safety phrases but few indications of effective means for disposal of products concerned and the precautions to be taken at disposal. (It is understood that information of this nature may well be included in the List in future.)

It is interesting to note the suggestion that certain substances should be banned from the workplace because they affect the aquatic environment adversely—to a degree out of all proportion to their hazard to people. (An example is hexabutyldistannoxone, which is harmful to marine invertebrates.) Perhaps it could be argued that a ban on such substances would be merely an extension of the prohibitions listed already in Schedule 2 of the Regulations.[4]

Conclusions

Preparing an integrated assessment is analogous to assessment under the Regulations in that it commences with identification of the substances concerned with acquisition of data as to their toxicology. However, in integrated risk assessment the toxicology should be not only that for human beings but also for other species—flora and fauna—which are likely to be exposed to contact with the materials, both in their original forms and the wastes derived from them. This toxicological information should include consideration of the routes of exposure. With exceptions such as food additives and pharmaceuticals, toxicology for human beings is deduced by experts from animal and genotoxicity studies.

If appropriate (it is happening already with new projects) an integrated risk assessment can take account of likely effects on public health and on the environment at large—including air, soil, and water—the best available quantification of the consequences of an activity for individuals and species. Such assessments form a basis for decisions in the construction and management of new units, to ensure adequate controls not only for internal purposes but also for the benefit of ecosystems nearby.

In essence, the concept could be (as summarized in the United Nations Environmental Programme's phrase) 'pollution prevention pays', which seems rather more reasonable and agreeable than the familiar tag about 'making the polluter pay'.

This chapter is concluded with two addenda, as follows:
Addendum I Three Case Studies in Integrated Assessment and Control
Addendum II Terms and Definitions Proposed.
(References precede the Addenda.)

References

1. J. Orme and E. V. Ohanian, 'Health Advisories for Pesticides' in 'Chemistry, Agriculture, and the Environment', ed. M. L. Richardson, The Royal Society of Chemistry, Cambridge, 1991, ch. 23, p 444.

2. M. J. Gardiner *et al.*, 'Results of a case-control study of leukaemia and lymphoma among people near Sellafield nuclear plant in West Cumbria'. *Brit. Med. J.*, **300** (February 1990).
3. 'Environmental Protection Act', HMSO, London, 1990.
4. 'Health and Safety The Control of Substances Hazardous to Health Regulations, SI 1657, 1988'. HMSO, London, 1988.
5. 'Approved Codes of Practice for Control of Substances Hazardous to Health', HMSO, London.
6. D. V. Parke *et al.*, 'Current Procedures for the Evaluation of Chemical Safety', in 'Risk Assessment of Chemicals in the Environment', ed. M. L. Richardson, The Royal Society of Chemistry, London, 1988 (reprinted 1990).
7. 'COSHH in Laboratories', The Royal Society of Chemistry, London, 1989.
8. J. L. Vosser, 'The European Community Chemicals Notification Scheme and Environmental Hazard Assessment', in 'Toxic Hazard Assessment of Chemicals', ed. M. L. Richardson, The Royal Society of Chemistry, London, 1986 (reprinted 1988, 1989).
9. 'Proposal for a Council Directive amending for the seventh time Directive 67/548/EEC on the approximation of the laws, regulations and administrative provisions relating to the classification, packaging and labelling of dangerous substances 90/C 33/03'. *Official Journal of the European Communities* No. C 33/3 Brussels, (13 February 1990).
10. E. Heinisch and S. Klein, 'Contamination of surface water, sediment, and biota by a pesticide manufacturing factory', in 'Chemistry, Agriculture, and the Environment', ed. M. L. Richardson, The Royal Society of Chemistry, Cambridge, 1991, ch. 16, p. 255.
11. E. Funari *et al.*, 'Herbicide groundwater contamination' in 'Chemistry, Agriculture, and the Environment', ed. M. L. Richardson, The Royal Society of Chemistry, Cambridge, 1991, ch. 14, p. 235.
12. A. Amadi, 'The effects of nitrogen fertilizer factory discharges to soil', in 'Chemistry, Agriculture, and the Environment', ed. M. L. Richardson, The Royal Society of Chemistry, Cambridge, 1991, ch. 13, p. 221.
13. 'IUPAC—Glossary of Terms in Chemical Toxicology and Ecotoxicity', ed. M. L. Richardson and J. H. Duffus, 2nd draft, 1990.

References Cited in Addendum II

'Acute and Sub-acute Toxicology', ed. V. K. Brown, Edward Arnold, London, 1988.

'Environmental Toxicology and Ecotoxicity', ed. J. H. Duffus, Environmental Health Series No. 10, World Health Organization Regional Office for Europe, Copenhagen, 1986.

'Glossary of Terms Related to Health, Exposure and Risk Assessment', (U.S.) Environmental Protection Agency EPA/450/3-88/016.

IARC monographs on the evaluation of carcinogenic risk to humans, Supplement 7. World Health Organization, International Agency for Research in Cancer, Lyon, 1987.

'International Programme on Chemical Safety Glossary of terms for use in IPCS Publications, 1987'.

'Integrated Risk Information System of the United States Environmental Protection Agency', EPA/600/6-86/032a.

'International Register of Potentially Toxic Substances (*sic*) English–Russian

Glossary of Selective Terms in Preventive Toxicology', Interim Document. United National Environment Programme, Moscow, 1982.
'Epidemiology: A Dictionary of Terms', ed. J. M. Last, International Epidemiological Association, Ottawa, c/o University of Ottawa, 1981.
'IUPAC—Glossary of Terms Used in Biotechnology for Chemists', ed. B. Nagel *et al.*, in press.
Risk Assessment of Chemicals in the Environment', ed. M. L. Richardson, The Royal Society of Chemistry, London, 1988.
Toxic Hazard Assessment of Chemicals, ed. M. L. Richardson, The Royal Society of Chemistry, London, 1986,
'WHO Basic Terminology for Risk and Health Impact Assessment (*sic*) and Management, Annex 3, 12 April 1988', World Health Organization, Geneva, 1988.

Addendum I Three Case Studies in Integrated Assessment and Control

1. The Use of Solvents at a Large Civil Engineering Work

In essence the unit concerned comprised a number of very large rams which were used to operate gates weighing upwards of 3000 tonnes.

Usually, the gates were operated remotely from an air-conditioned control room but they also could be controlled from small compartments which were not well-ventilated. The whole construction was near an important waterway, and a total of some 70 000 electrical relays and contacts was involved.

Prior to assessment for COSHH, the electrical contacts were cleaned using 1,1,2-trichloroethene (shown in EH 40 as trichloroethylene) as solvent of choice, often sprayed from aerosol cans which contained propane–butane mixtures or chlorofluorocarbons as propellants. It was evident that in some situations the occupational exposure limit for the solvent and/or propellant would be exceeded.

At an early stage in the assessment the use of other solvents was considered, including dichloromethane. The latter offered the advantage of being less hazardous but because of the relatively high temperatures experienced in some of the control rooms (and bearing in mind its low boiling point and high volatility) this was rejected quickly. Consideration was given to halogenated solvents with higher boiling points but those which were considered satisfactory from the point of view of the electrical engineers were found either to leave 'gummy' deposits or to be deficient in data on the toxicology of inhalation. The use of replacement solvents such as these would have required breathing apparatus, a considerable encumbrance in the locations concerned.

An entirely different approach was recommended, under which electrical contact units were to be replaced at intervals, or upon failure, with spares and the originals returned to a central workshop where they could be cleaned in small baths of 1,1,2-trichloroethene. The workshop was large and very well-ventilated, and the cleaning could be carried out under controlled conditions. Further, there was no need to use aerosols, so the risks associated with chlorofluorocarbons and highly-flammable propellants in confined spaces were eliminated, and also their undesirable effects to the air environment removed entirely.

The technique of replacement proved very successful and this was judged to be a distinctly beneficial assessment and solution to the problem.

2. Discharges to Aquatic Environment of Dyestuffs Based on Benzidine Congeners

In the course of dyeing a multicoloured fabric the dyer was found to be discharging some fifty per cent of a mixture of dyestuffs.

The trade effluent from the factory was highly coloured; from time to time the discharge from the sewage treatment works also was coloured—leading to coloration of the receiving river, complaints from anglers, and from others wishing to enjoy riverside amenities.

The dyestuffs concerned were examined in detail and some of them were found to be based on benzidine and its congeners; such dyestuffs carried the risk phrase 'R–45'. Thus the management of the factory was faced with the possibility of harm to employees, as well as the consequences of polluting the river.

The suppliers were approached and the materials based on benzidine and its congeners were replaced with others more acceptable from the occupational and environmental points of view. In addition, some persistent dyestuffs which were responsible for the problem of coloration of the effluent were replaced with others which were retained better on the fabric or were absorbed more easily into sewage sludge. Briefly, some relatively cheap dyestuffs were replaced by others which were more expensive in unit terms but more effective in dyeing the fabric and generally more acceptable otherwise—especially in respect of discharges to the aquatic environment; overall, the cost of the work was reduced.

3. A Pharmaceutical Intermediate

The manufacture of a pharmaceutical included as an intermediate stage the synthesis of an *N*-nitroso compound, with an associated explosion hazard in addition to the mammalian toxicity: there was also at this stage of synthesis discharge of a substantial quantity of nitrite. (In the course of investigation, significant risks of human exposure were found to arise from the nitrite.)

It was decided that the synthesis would be carried on in future entirely within a containment area, under comprehensive control by computer and with remote handling of materials. Since the area could not be entered during normal operations, fire and explosion hazards were eliminated and there could be no risks to operators from nitrosamines. Pilot-scale experiments disclosed a significant wastage of nitrite, and computer control of the process helped to reduce this: further, under the new arrangements, any excess nitrite was destroyed chemically.

This example indicated the benefits that can arise from considering all aspects of a synthesis together, including the effluents and other wastes as well as the hazards associated immediately with the product or intermediate concerned.

Addendum II Terms and Definitions Proposed

As in all sciences it is desirable when undertaking assessments for the Regulations and for integrated purposes that the meanings of the terms used should be clear and apparent to all concerned. With a view to the resolution of problems in terminology the Commission on Toxicology within the Division of Clinical Chemistry of the International Union of Pure and Applied Chemistry is compiling a 'Glossary of Terms on Chemical Toxicology and Ecotoxicology'. Some definitions from among the proposals under consideration are quoted in the pages following. IUPAC recommendations will be published in due course (it is anticipated in 1992, in the journal *Pure and Applied Chemistry*), and may well be different in their final form from the current proposals.*

Before going on to the selection in alphabetical order from the definitions proposed, 'Hazard', 'Risk', and 'Safety' are expressed as follows:

Hazard. The set of inherent properties of a chemical, a physical substance, or of a mixture of substances under conditions of production, use, or disposal, that would be capable of causing adverse effects in man or the environment; alternatively, any biological process depending upon the particular degree of exposure to those properties.

Risk. The likelihood of suffering harmful effects from exposure to hazard as defined above. Usually it is expressed as a probability of the occurrence of an adverse effect (as an example, the ratio anticipated between the number of individuals experiencing an adverse effect within a given time and the total number of individuals exposed).

Safety. The practical certainty that, under defined conditions, injury will not result from a hazard; alternatively, the high probability that under given conditions for the quantity and manner of use of a substance no injury will result from it.

As noted, the terms and definitions below were derived from a glossary prepared by a working party of the International Union of Pure and Applied Chemistry, Clinical Chemistry Division (see References).

Acceptable risk. The probability of suffering disease or injury that is deemed acceptable by an individual, group, or society. Acceptability of risk depends on scientific data, together with social, economic, environmental, and political factors, and on the perceived benefits and ethical issues related to the use of a chemical or process. *J. H. Duffus 1986*

Adverse effect. An abnormal, undesirable, or harmful effect to an organism,

*IUPAC nomenclature and definitions are the copyright of IUPAC. It is understood that a liberal policy is followed with regard to giving permission to reproduce such material and the draft definitions here are quoted by permission.

indicated by some result such as mortality, altered food consumption, altered body and organ weights, altered biochemical activities or visible or microscopic pathological changes. *modified from J. H. Duffus 1986*

Note: A statistically significant change from the normal state of an organism exposed to a chemical is not *necessarily* a biologically adverse effect. The magnitude of departure from the normal range, the consistency of the out-of-range effect, and the relationship of the effect to the physiological, biochemical, and total well-being of the test organism have to be considered. An effect may be considered adverse if it causes functional or anatomical damage, causes irreversible change in the homeostasis of the organism, increases the susceptibility of the organism to other chemical or biological stress, or reduces an organism's ability to respond to an additional environmental challenge. A non-adverse effect usually will be reversed when exposure to the chemical ceases.

Attributable risk. The difference between the risk of exhibiting a certain adverse effect in the presence of a toxic substance and that risk in the absence of the substance. *IRIS, 1986*

Attributable risk (among exposed). The proportion of the risk among the exposed that is attributable to the exposure. It is computed as:

$$AR_e = \frac{I_e - I_u}{I_e} = \frac{RR - 1}{RR}$$

where I_e is incidence rate among those exposed
I_u is incidence rate among those unexposed
RR is risk ratio, or I_e/I_u. *J. M. Last, 1981*

Attributable risk per cent (exposed). Attributable risk per cent ($ARe\%$) is the attributable risk expressed as a percentage (rate) rather than as a proportion. It is computed as:

$$ARe\% = \frac{I_e - I_u}{I_e} \times 100$$

where I_e is incidence rate among those exposed
I_u is incidence rate among those unexposed *J. M.Last, 1981*

Benefit. A gain to a population. Expected benefit incorporates an estimate of the probability of achieving a gain. *M. L. Richardson, THAC, 1986*

Carcinogen. An agent (whether chemical, physical, or biological) that is capable of increasing the incidence of malignant neoplasms. The induction of benign neoplasms in some circumstances may contribute to the judgment that an agent is carcinogenic. *IARC, 1987*

Databank. Pre-selected factual information in summary form, with a sophisticated search system to enable the information required to be located.
modified from M. L. Richardson, THAC, 1986

Database. Usually on-line computer-based information, retrieved files containing information such as bibliographies, references, and in some cases abstracts of papers, with the more recent literature in a particular field.
modified from M. L. Richardson, THAC, 1986

Detoxification.

(i) A process, or processes, of metabolism which renders a toxic molecule less toxic by removal, alternation, or masking of active functional groups.

(ii) To treat patients suffering from poisoning in such a way as to reduce the probability and/or the severity of harmful effects.

M. L. Richardson, RACE, 1988

Dose. The amount of a substance administered to, or absorbed by, an organism. 'Uptake' is preferred to the usual term 'dose', because the precise dose administered is very difficult to measure and because 'uptake' indicates effective exposure. *modified from IRPTC, 1982*

Ecology. The branch of biology that studies the interactions between living organisms and all other factors (including other organisms) in their environment. Such interactions include factors that determine the distributions of living organisms. *IPCS, 1987*

Ecotoxicology. The study of toxic effects of chemical and physical agents on living organisms as well as human beings, especially on populations and communities within defined ecosystems; it includes transfer pathways of these agents and their interaction with the environment. (See also: Toxicology, ecological) *M. L. Richardson, RACE, 1988*

Environment. The aggregate, at a given moment, of all external conditions and influences to which a system is subjected. The term 'system' covers all living organisms, including human beings. *IPCS, 1987*

Environment fate. The density of a chemical or biological pollutant after release into the environment. *EPA, 1989*

Note: It involves temporal and spatial considerations of transport, transfer, storage, and transformation.

Environmental protection.

(i) A complex of measures including monitoring of environmental pollution, the development and practice of environmental protection measures (legal, technical, hygienic), and including: risk assessment, risk management, and risk communication. *modified from IRPTC, 1982*

(ii) Actions taken to prevent or to minimize adverse effects to the natural environment.

Environmental quality objective ('EQO'). The quality to be aimed for in a particular aspect of the environment, for example, 'the quality of water in a river such that coarse fish can maintain healthy populations'. Unlike an environmental quality standard the EQO usually is not expressed in quantitative terms. *M. L. Richardson, RACE, 1988*

Environmental quality standard ('EQS'). The concentration of a potentially toxic substance that can be allowed in an environmental component, usually air (air quality standard), or water, over a defined period. Synonym: Ambient standard. (See: Limit value) *M. L. Richardson, RACE, 1988*

Environmental transformation. The conversion of a chemical into a derivative by physicochemical processes occurring in the natural environment.

Epidemiology. The study of health factors affecting the occurrence and resolution of disease and other health-related conditions or events in populations. *J. M. Last, 1981*

Exposure. The amount and the physical conditions of interaction between organisms and toxicants; the process by which a toxic substance is introduced into or absorbed by the organism (or population) by any route. 'Exposure' generally is more descriptive, hence often more meaningful, than 'dose' but the two words frequently are used interchangeably.

modified from V. K. Brown, 1988, and IPCS, 1987

Note: The concentration (or intensity) of a particular physical or chemical agent that reaches the target, usually expressed in numerical terms of duration, frequency, and concentration (for chemical agents and micro-organisms) or intensity (for physical agents such as radiation).

Harm. Damage to a species or individual. In the case of a toxicant this is a function of the concentration to which the organism is exposed and of the time of exposure. *modified from M. L. Richardson, RACE, 1988*

Harmful substance. A material which, following contact with a human organism (under the working conditions of everyday life) can cause disease, health variations, or other adverse effects—either at the time or in later periods of life of the present or future generations. Synonym: Noxious substance. *modified from IRPTC, 1982*

Individual risk. The probability that an individual person will experience an adverse effect. It is identical to Population risk unless specific sub-groups of population can be identified that have different (higher or lower) risks.

IRIS, 1986

The risk to any particular member of a population, either a worker or a member of the public, including those living within a defined radius from an incident; alternatively, to individuals who follow a particular pattern of life.

Limit value ('LV'). The limit at or below which member states of the European Community must set their environmental quality and emission standards. Such limits are given by Community Directives.

M. L. Richardson, RACE, 1988

Maximum permissible daily dose. The maximum daily dose of substance the penetration of which into the human body during a lifetime will not cause diseases or health hazards that can be detected by current methods of investigation, or will not affect adversely future generations.

modified from IRPTC, 1982

Permissible exposure limit. The level, usually a combination of time and concentration, at which humans may be exposed safely to chemical or physical agents in their immediate environment.

Pollutant. Any undesirable solid, liquid, or gaseous matter in a gaseous, liquid, or solid medium. A 'Primary pollutant' is a pollutant emitted into the atmosphere, sediments, soil, or water from an identifiable source; a 'Secondary pollutant' is one formed by chemical reaction in the atmosphere, sediments, soil, or water. *modified from IPCS, 1987*

Note: Any factor which affects the environment adversely.

Pollution. Any undesirable modification of air, sediments, soil, water, or of

food materials by a substance or substances solid, liquid, or gaseous in form that is or are toxic, may have adverse effects on health, is or are offensive though not necessarily harmful to health. *modified from J. M. Last, 1981*
Note: The introduction of pollutants into the environment.

Population at risk. The number of persons who could develop the adverse health effect under study and may be exposed to the risk factor of interest.
Note: It also can mean the group of the working population likely to be affected most by a given exposure or that is exposed to high concentrations from which adverse effects might follow. For example, all persons in a population who have not developed immunity to an infectious disease are, if exposed to it, at risk of developing the disease. Conversely, persons who have already the disease concerned are excluded, in studies of its incidence, from the Population at risk.

Relative risk. Ratio of the cumulative incidence of those exposed to a factor to the incidence rate of those not exposed. Synonyms: Cumulative incidence ratio, Risk ratio. *modified from J. M. Last, 1981*
Note: (i) Prospective study yields the data needed to calculate incidence rates. These in themselves are an estimate of the risk that an individual will get the condition of interest. The ratio of these risk estimates is the Risk ratio or Relative risk.
(ii) The ratio of the risk of disease or death among the exposed population to the risk among the unexposed; this usage is synonymous with Risk ratio.
(iii) Alternatively, the ratio of the cumulative incidence rate in the unexposed—that is, the Cumulative incidence ratio.
(iv) The term Relative risk also has been used synonymously with 'odds ratio' and, in some biostatistical articles, used for the ratio of forces of morbidity. The use of the term Relative risk for several different quantities arises from the fact that for 'rare' diseases (such as most types of cancer) the quantities all approximate to each other. For relatively common occurrences (such as neonatal mortality in infants of under 1500 g weight at birth) the approximations do not hold.

Risk aversion. A term used to describe the attitude of most people towards risk. Used correctly it refers to one's willingness to pay a premium above the expected value to avoid a gamble (that is, people wish to avoid risks to their health).

Risk estimation. (*cf*: Risk evaluation). The quantification of dose-effect and dose-response relationships for a given chemical substance, showing the probability and the nature of a physiological effect of exposure to the substance.
Comment: Note is to be taken of the quantification of dose-effect and dose-response relationships for a given environmental agent, showing the probability and nature in a general scientific sense of the health effects of exposure to the agent.

Risk evaluation. The qualitative or quantitative relationship between risks and benefits.
Comment. It involves the complex process of determining the significance or

value of the identified hazards and estimated risks to those concerned with or affected by the decision. Therefore it includes the study of Risk perception and striking a balance between risks perceived and benefits perceived. It can be expressed also as the comparison of calculated risks or the effect on public health of exposure to an environmental agent, with risks caused by other agents or by societal factors, and with the benefits associated with the agent—as a basis for a decision about Acceptable risk.

Risk perception. An integral part of Risk evaluation. The subjective perception of the gravity or importance of the risk based on the subject's knowledge of different risks and the moral and political judgment of their importance. Risk perception also is of importance at the stage of hazard identification. *WHO, 1988*

Short-term exposure limit ('STEL'). The administrative time-weighted average ('TWA') airborne concentration to which employees may be exposed for periods up to fifteen minutes with no more than four such excursions per day and at least sixty minutes between them. *IRIS, 1986*

Sub-acute. A term used to describe a form of exposure being not so short as Acute but not long enough to be called 'long-term' or 'chronic'. An imprecise term used to describe exposures of intermediate duration.

V. K. Brown, 1988

Note: Sometimes called 'sub-chronic'.

Sub-chronic effect. A biological change resulting from change in the environment lasting for about ten percent of the lifetime of the test organism.

modified from IRIS, 1986

Note: In practice such an effect usually is identified as resulting from multiple or continuous exposures occurring over three months.

Threshold limit values ('TLV'). Permissible exposure limits for occupational exposure to airborne contaminants, published by the American Conference of Government Industrial Hygienists (ACGIH). The values represent the average concentration (in mg m^{-3}) for an eight-hour working day and a forty-hour week to which nearly all employees may be exposed repeatedly, day after day, without adverse effect. *modified from IRIS, 1986*

Time-weighted average ('TWA'). A term applied to the expression of permissible levels for occupational exposure. Time-weighted averages permit exposure above a level provided that in the course of a working day or shift it is compensated by equivalent excursions below the level. In some national lists, the magnitude, duration, and frequency of permissible excursions are specified. *IRPTC, 1982*

Time-weighted average concentration ('TWAC'). The concentration of a substance to which a person is exposed in ambient air, averaged over a period, usually of eight-hours. For example, if a person is exposed to 0.1 mg m^{-3} for six hours and 0.2 mg m^{-3} for two, the eight-hour TWA is:

$$\frac{(0.1 \times 6 + 0.2 \times 2)}{8} = 0.125 \text{ mg m}^{-3}$$ *J. H. Duffus, 1986*

Tolerance. An adaptive state characterized by diminished effects of the same dose of a material.

modified from J. H. Duffus, 1986; IPCS, 1985; and M. L. Richardson, RACE, 1988
Comment: The process leading to tolerance is called 'adaptation'. In food toxicology it is the dose that an individual can receive without showing an effect. It is also the ability to experience exposure to possibly harmful amounts of a substance without showing an adverse effect and the ability of an organism to survive in the presence of a toxic substance. Increased tolerance may be acquired by adaptation to constant exposure.
Toxic*. Able to cause injury to living organisms as a result of chemical interaction; producing toxicity. *J. H. Duffus, 1986*
Toxicity. Any harmful effect of a chemical or a drug on a target organism. See also: Acute, Chronic, and Sub-chronic toxicity.
Note: It is: (i) The capacity to cause injury to a living organism. A highly toxic substance will cause damage to an organism if administered in very small amounts, and a substance of low toxicity will not produce an effect unless the amount is very large. However, toxicity cannot be defined in quantitative terms without reference to the quantity of substance administered or absorbed, the way in which this quantity is administered (as examples, by inhalation, ingestion, or injection) and distributed in time (as examples, in single or repeated doses), the type and severity of injury, and the time needed to produce the injury.
(ii) Any adverse effects of a chemical on a living organism (see: Adverse effect). The term also is used to describe the ability of a chemical to cause adverse effects. The degree of toxicity produced by any chemical is directly proportional to the concentration of exposure and the time of exposure. This relationship varies with the stage of development of the organism and with other factors, such as nutritional status.
(a) Acute toxicity. Adverse effects occurring within a short time of exposure to a single dose of a chemical, or to multiple doses over 24 hours or less.
(b) Sub-acute (sub-chronic) toxicity. Adverse effects occurring as a result of repeated daily exposure to a chemical for part of the organism's lifespan (usually not exceeding 10%). With experimental animals, the period of exposure may range from a few days to six months.
(c) Chronic toxicity. Adverse effects occurring in a living organism as a result of a repeated daily exposure to a chemical for a large part of its life span (usually more than 10%). With experimental animals, this usually means a period of exposure of more than three months.
(iii) A measure of incompatibility of a substance with life. This quantity is the reciprocal of the absolute value of dose ($1/LD_{50}$) or concentration ($1/LC_{50}$). *modified from J. H. Duffus, 1986*
Toxicology. The scientific discipline involving the study of the actual or possible danger presented by the harmful effects of substances (poisons) on living organisms and ecosystems; of the action, mechanisms, diagnosis, prevention, and treatment of intoxications. *modified from IRPTC, 1982*

*The Concise Oxford Dictionary gives: 'Of poison; . . . poisonous', 'toxicity' and: '*toxin*, n. A poison.'

Toxicology, ecological. (Synonym: Ecotoxicology, *q.v.*) A branch of toxicology studying the effects of substances which might be toxic on the flora and fauna, ecosystems, and the transport of harmful substances in the biosphere, especially in food chains. *IRPTC, 1982*

Toxification. The metabolic conversion of a possibly toxic substance into one more toxic. *modified from V. K. Brown, 1988*

Toxin. A toxic organic substance produced by the normal metabolism of a living organism. *modified from M. L. Richardson, RACE, 1988*

Toxinology. Toxicology of toxins. *modified from V. K. Brown, 1988*

Waste. Anything that is discarded or otherwise dealt with.

Note: It also can be regarded as a commodity of negative value. Wastes include:

(i) Solid waste. This includes all wastes that are not liquid effluents in bulk, discharged to sewer or a receiving water, or gases (with entrained solids) discharged to atmosphere.

(ii) Controlled waste. This comprises household, industrial, and commercial waste.

(iii) Municipal waste. The aggregate of all those wastes that the collection authorities have a duty to collect, whether they do so themselves or employ an agent to collect the waste.

Comment: Therefore, municipal waste can be a mixture of household, commercial, civic amenity, street cleaning, privy and cesspool wastes, wastes from maintenance and cleaning of public parks and open spaces, ash from municipal incinerators, waste from municipal buildings, construction and demolition works, and from municipal undertakings such as transport, abandoned motor vehicles, *etc.*

(iv) Civic amenities waste. Waste arising in households that for some reason is not collected free of charge by the collection authority.

Note: Typically it consists of garden waste, from building and maintenance work on a 'do-it-yourself' basis, from car maintenance, items of unwanted furniture, and so forth. It should include no industrial or commercial waste but can (and often does) contain (as examples): household chemicals, oils, paint, and pesticides.

(v) Hazardous waste. Those wastes not indicated on a listing of non-hazardous wastes.

Note: It includes wastes that pollute water but have little effect otherwise on health or the environment. Called sometimes 'notifiable waste'.

(vi) Special waste. Waste capable of causing harm or injury to people or to organisms directly exposed to it.

Comment: This is not the same as being harmful environmentally. The waste may or may not be damaging environmentally but its acceptance at a specific disposal site will depend on the attributes of the site at least as much as those of the waste.

(vii) Toxic or poisonous waste. Waste where human toxicity is the primary hazard.

Xenobiotic. A chemical not a natural component of the living organism exposed to it. Synonyms: Foreign substance or compound, exogenous or anthropogenic material. *modified from B. Nagel,* et al., *1989*

Appendix: Signs and Labels

The basis for marking and labelling hazardous substances is set forth in SI No. 1244, the Regulations for 'Classification, Packaging and Labelling of Dangerous Substances, 1984'. These Regulations are supplemented by a total of eight schedules which classify substances, symbols, and labels for a variety of purposes, and also by substantial lists of substances, classifications and requirements issued by the Health and Safety Executive—and by other organizations, such as the International Air Transport Association, International Civil Aviation Organization, and International Maritime Organization.

Table XVII, page 167 below is an attempt by the editors to present the main mandatory signs and labels in a convenient form: it should not be assumed that it necessarily reflects in any particular case the actual requirements of the Regulations; reference always should be made to the Regulations concerned.

Signs

Signs covering safety measures are mandatory in certain circumstances: usually they comprise a relevant blue and white graphic device (gloves, helmet, *etc.*) and words such as:

Eye protection must be worn
Guards must be used
Hand protection must be worn
Masks must be worn
Protective garments must be worn
Protective footwear must be worn
Respirators must be worn
Safety helmets must be worn
Use ear protectors
Wear face shield
Wear goggles
Wear safety harness.

TABLE XVII *Hazard Phrases, Symbols, and Signs*

Word or phrase	*Symbol*	*Colour of sign**
COMPRESSED GAS	Gas cylinder	Green
CORROSIVE	Drops of liquid falling on surface or hand	Upper triangle (containing symbol): white Lower triangle (containing word or phrase): black
DANGEROUS SUBSTANCE	Exclamation mark	White
DANGEROUS WHEN WET	Flame and smoke	Red
EXPLOSIVE	Matter scattering	Red
EXTREMELY FLAMMABLE, or HIGHLY FLAMMABLE	Flame and smoke	Red
FLAMMABLE GAS, or FLAMMABLE LIQUID	Flame and smoke	Red
FLAMMABLE SOLID	Flame and smoke	Alternating red and white vertical bars
HARMFUL	Heavy St. Andrew's Cross	Orange
IRRITANT	Heavy St. Andrew's Cross	Orange
OXIDIZING, OXIDIZING AGENT, or ORGANIC PEROXIDE	Circular device emitting flame and smoke	Yellow
SPONTANEOUSLY COMBUSTIBLE	Flame and smoke	Upper triangle (containing symbol): white Lower triangle: red
TOXIC, TOXIC GAS, or VERY TOXIC	Skull and cross-bones	White
RADIOACTIVE	'Windmill'	Upper triangle (containing symbol): yellow Lower triangle (containing additional information): white

*The signs are diamond-shaped—some divided into upper and lower triangles by a horizontal line (in general in such cases the symbol appears in the upper triangle and the word or phrase below).

Bibliography

(The references are arranged in alphabetical order, so far as possible by surname of the author or editor first given: if an author or editor is not stated (or unknown), the reference is identified by title.)

Advisory Committee on Dangerous Pathogens (see: *Categorization of Pathogens*).

M. R. Alderson, 'Occupational Cancer', Butterworths, London, 1986.

'Amphibole Asbestos', Cape Asbestos, London, n.d.

'Approved Codes of Practice on Control of Substances Hazardous to Health', HMSO, London, various dates.

Control of Carcinogenic Substances.

For Potteries.

Fumigation Operations.

General.

Vinyl Chloride at Work.

Asbestosis Research Council (see: *Technical Notes*)

Second International Meeting on Chemical Sensors, Bordeaux, France, ed. J.-L. Aucouturier *et al.*, University of Bordeaux, 1986.

A. Bailey and P. A. Hollingdale-Smith, 'A Personal Diffusive Sampler for evaluating time-weighted exposure to organic gases and vapours', *Ann. Occup. Hyg.*, 1977, **20**, 345.

A. Brown, R. H. Brown, and K. J. Saunders, 'Diffusive Sampling An Alternative Approach to Workplace Air Monitoring', The Royal Society of Chemistry, London, 1987.

'Biosafety in Microbiological and Biomedical Laboratories', U.S. Government Printing Office, Washington, 1988.

'Methods for Assessing and Reducing Injury from Chemical Accidents', ed. P. Bourdeau and G. Green, John Wiley & Sons, Chichester, 1989.

'Handbook of Reactive Chemical Hazards', ed. L. Bretherick, Butterworths, London, 1985.

'Hazards in the Chemical Laboratory', ed. L. Bretherick, The Royal Society of Chemistry, London, 1986.

'British National Formulary', British Medical Association and Royal Pharmaceutical Society of Great Britain, London, various dates.

'British Pharmaceutical Codex', The Pharmaceutical Press, London, various dates.

'British Pharmacopoeia', HMSO, London, various dates.

'British Standards', British Standards Institution, Milton Keynes, dates as shown below.

BS 873: (various parts) 1980–1987 Road traffic signs and internally illuminated bollards.

BS 3406: 1985 Methods for the determination of particle size of powders, Part 4: Optical microscope method.
BS 4803 Part 2: 1983 Radiation safety of laser products and systems
BS 5343: 1976 Gas detector tubes.
BS 5378: (various parts) 1980–1982 Safety signs and colours.
BS 5655 Part 6: 1985 Lifts and service lifts.
BS 5656: 1983 Safety rules for the construction and installation of escalators and passenger conveyors.
BS 5750: (various parts) 1981–1987 Quality systems.
BS 6069 Sec 3.1: 1989 (ISO 876: 1988) (Air quality) Workplace atmospheres method for the determination of vinyl chloride using a charcoal tube and a gas chromatograph.
BS 7229: 1989 British Standard Guide to quality systems auditing.

R. H. Brown, J. Charlton, and K. Saunders, 'The Development of an Improved Diffusive Sampler', *Am. Ind. Hyg. Ass. J.*, 1981, **42**, 865.

V. K. Brown, 'Acute Toxicity in Theory and Practice', John Wiley, Chichester, 1980.

Acute and Sub-acute Toxicology, ed. V. K. Brown, Edward Arnold, London, 1988.

E. Browning, 'Toxicity of Industrial Solvents', HMSO, London, 1953.
ibid. See: R. Snyder (ed.).

'The Merck Index, 11th (Centennial) edition', ed. S. Budavari, Merck & Co., N. J. Rahway, 1989.

S. Burnouf, 'Annual Health Screening of printing production workers', Unpublished, 1989.

S. Burnouf, 'Pre-employment health considerations for newspaper production workers', Unpublished, 1988.

'Business Monitor: Retail', HMSO, London.

Casarett and Doull, 'Toxicology: the Basic Science of Poisons', Macmillan, New York, 1986.

'Categorization of Pathogens According to Hazard and Categories of Containment', HMSO, London, 1984.

S. S. Chissick and R. Derricott, 'Occupational Health and Safety Management', John Wiley & Sons, Chichester, 1981.

S. S. Chissick and R. Derricott, 'Asbestos. Vol. II Properties, Application and Hazards', John Wiley & Sons, Chichester, 1983.

K. Y. K. Chung and N. P. Vaughan, 'Comparative laboratory trials of two portable direct-reading dust monitors', *Ann. Occup. Hyg.* 1989, **33**, 591.

Cmd. 728 'Measuring up to the Competition', HMSO, London, 1989.

C. H. Collins, 'Laboratory Acquired Infections: History, Incidence, Causes and Prevention', Butterworths, London, 1988.

C. H. Collins and J. M. Grange, 'The Microbiological Hazards of Occupations', Science Reviews, Leeds, 1990.

Commission of the European Communities [see also: K. Rasmussen, (ed.)]. Community Documentation Centre on Industrial Risk Volume 3 (Consolidated Volume containing also context of Vol. 1 and Vol. 2) Ispra: Commission of the European Communities, 1990.

Commission of the European Communities, Council Directive of 16 December 1988 amending Directive 80/1107/EEC. *Official Journal of the European Communities*, No. 1356/74 to L356/78, Brussels 24 December 1988.

Commission of the European Communities, proposal for a Council Directive amending for the seventh time Directive 67/548/EEC on the approximation of the laws, regulations and administrative provisions relating to the classification, packaging and labelling of dangerous substances 90/C 33003. *Official Journal of the European Communities*, No. C 33/3, Brussels 13 February 1990.

'Van Nostrand's Scientific Encyclopaedia', 7th Ed, ed. D. Considine, Van Nostrand Reinhold, New York, 1989.

'Controlling Airborne Contaminants in the Workplace', British Occupational Hygiene Society Technical Guide No. 7. Northwood: Science Reviews, 1988.

P. Cooper, 'Poisoning by Drugs and Chemicals', Alchemist Publications, London, 1974.

'COSHH: An open learning course', Health and Safety Executive, Bootle, n.d.

'COSHH: Assessment', HMSO, London, n.d.

'COSHH Assessment Workbook', Institution of Occupational Safety and Health.

'COSHH Guidance for the Construction Industry', HMSO, London, n.d.

'COSHH Guidance for Universities, Polytechnics and Colleges of Further Education', HMSO, London, n.d.

'COSHH in Laboratories', The Royal Society of Chemistry, London, 1989.

'COSHH Regulations 1988: Register', Progressive Risk Assessments Limited, 17 Church Road, Northfield, Birmingham, B31 2JZ.

'COSHH—your questions answered', *Chemistry in Britain*, 1989 **25**, pp 1188–1189.

N. P. Crawford and A. J. Cowie, 'Quality Control of Airborne Asbestos Fibre Counts in the U.K.—the Present Position', *Ann. Occup. Hyg.* 1984, **28**, 391.

'Detection and Measurement of Hazardous Gases', ed. C. F. Cullis and J. G. Firth, London, Heinemann, 1981.

'Toxicology of Drugs and Chemicals', W. B. Deichmann and H. W. Gerarde, Academic Press, New York, 1969.

Department of Environment (see: 'Reports on Public Health').

Department of Health and Social Security (see: 'Reports on Public Health', and 'Statutory Instruments').

'Directory of Members & Services', The Association of Consulting Scientists, London, various dates.

'Documentation of NIOSH Validation Tests', U.S. Department of Health and Human Services, Washington, D.C., 1977.

R. H. Driesbach,'Handbook of Poisoning', Blackwell, Oxford, 1977.

'Environmental Toxicology and Ecotoxicity', ed. J. H. Duffus, Environmental Health Series No. 10, World Health Organization Regional Office for Europe, Copenhagen, 1986.

'Environmental Protection Bill', HMSO, London, 1989.

S. Friess, 'History of Risk Assessment in Pharmacokinetics, in Risk Assessment: Drinking Water and Health, Vol. 8', National Academy of Science, Washington, D.C., 1987.

M. J. Gardiner *et al.*, 'Results of a case-control study of leukaemia and lymphoma among people near Sellafield nuclear plant in West Cumbria'; London: *British Medical Journal*, **300** (February 1990).

'Glossary of terms for use in IPCS Publications', International Programme on Chemical Safety, 1987.

'Glossary of Terms Related to Health, Exposure and Risk Assessment', (U.S.) Environmental Protection Agency EPA/450/3-88/016.

'The Pharmacological Basis of Therapeutics', ed. L. S. Goodman and A. G. Gilman, Macmillan, New York, 1985.

R. E. Gosselin, H. C. Hodge, R. P. Smith and M. N. Gleason, 'Clinical Toxicology of Commercial Products', Williams & Wilkins, Baltimore, 1976.

'Chromatographic Analysis of the Environment', ed. R. L. Grob, Marcel Dekker, New York, 1983.

'Guidance Notes'. (Health and Safety Executive) HMSO, London, various dates.

Environmental hygiene:

EH 1 Cadmium—health and safety precautions.
EH 2 Chromium—health and safety precautions.
EH 4 Aniline—health and safety precautions.
EH 5 Trichloroethylene—health and safety precautions.
EH 7 Petroleum based adhesives in building operations.
EH 8 Arsenic—Toxic hazards and precautions.
EH 9 Spraying of highly flammable liquids.
EH 10 Asbestos—exposure limits and measurement of airborne dust concentrations.
EH 11 Arsine—health and safety precautions.
EH 12 Stibine—health and safety precautions.
EH 13 Beryllium—health and safety precautions.
EH 16 Isocyanates: toxic hazards and precautions.
EH 17 Mercury—health and safety precautions.
EH 19 Antimony—health and safety precautions.
EH 20 Phosphine—health and safety precautions.
EH 21 Carbon dust—health and safety precautions.
EH 22 Ventilation of buildings: fresh air requirements.
EH 23 Anthrax: health hazards.
EH 24 Dust accidents in malthouses.
EH 25 Cotton dust sampling.
EH 26 Occupational skin diseases: health and safety precautions.
EH 27 Acrylonitrile: personal protective equipment.
EH 28 Control of lead: air sampling techniques and strategies.
EH 29 Control of lead: outside workers.
EH 31 Control of exposure to polyvinyl chloride dust.
EH 32 Control of exposure to talc dust.
EH 33 Atmospheric pollution in car parks.
EH 34 Benzidine based dyes. Health and safety precautions.
EH 35 Probable asbestos dust concentrations at construction processes.
EH 36 Work with asbestos cement.
EH 37 Work with asbestos insulating board.
EH 38 Ozone: health hazards and precautionary measures.
EH 40/90 Occupational exposure limits 1990.
EH 41 Respiratory protective equipment for use against asbestos.
EH 42 Monitoring strategies for toxic substances.
EH 43 Carbon monoxide.
EH 44 Dust in the workplace: general principles of protection.
EH 45 Carbon disulphide: control of exposure in the viscose industry.
EH 46 Exposure to mineral wools.
EH 47 Provision, use and maintenance of hygiene facilities for work with asbestos insulation and coatings.
EH 48 Legionnaires' disease.
EH 49 Nitrosamines in synthetic metal cutting and grinding fluids.

EH 50 Training operatives and supervisors for work with asbestos insulation coating.
EH 51 Enclosures provided for work with asbestos insulation and coating.
EH 52 Removal techniques for asbestos insulation coatings and insulation.
EH 53 Respiratory protective equipment for use against ionizing radiations.
EH 54 Assessment of exposure to welding fume.
Medical series:
MS 4 Organic dust surveys.
MS 5 Lung function.
MS 6 Chest *X*-rays in dust diseases.
MS 7 Colour vision.
MS 8 Isocyanates: medical surveillance.
MS 9 Byssinosis.
MS 10 Beat conditions, tenosynovitis.
MS 12 Mercury—medical surveillance.
MS 13 Asbestos.
MS 15 Welding.
MS 16 Training of offshore sick-bay attendants ('rig-medics').
MS 17 Biological monitoring of workers exposed to organo-phosphorus and carbamate pesticides.
MS 20 Pre-employment health screening.
MS 21 Precautions for the safe handling of cytotoxic drugs.
MS 22 Medical monitoring of workers exposed to platinum salts.
MS 23 Health aspects of job placement and rehabilitation.
'Guide for Consultancy', The Royal Society of Chemistry, London, 1981.
N. M. Harris, L. G. Shallenberger, B. S. Donaleski and E. A. Sales, 'A Retrospective Mortality Study of Workers in Three Major U.S. Refineries and Chemical Plants', *Journal of Occupational Medicine*, 1985: I. **27**, pp 283–292; II. **27**, pp 361–369.
J. M. Harrison and F. S. Gill, 'Occupational Health', Blackwell, London, 1987.
'Handbook of Occupational Hygiene', ed. B. Harvey *et al.*, Croner Publications, New Malden, 1990.
'Molecular Aspects of Toxicology', D. E. Hathway, The Royal Society of Chemistry, London, 1984.
'Pesticides Studied in Man', W. J. Hayes, Williams & Wilkins, Baltimore, 1982.
'Hazard and Risk Explained', HMSO, London, n.d.
'Hazard Data Sheets', Merck (formerly BDH), Poole, various dates.
'Health and Safety at Work Act', HMSO, London, 1974.
Health and Safety Executive (see: 'Approved Codes of Practice', 'Guidance Notes', 'Methods for Determination of Hazardous Substances', *etc.*).
'Health and Safety Statistics', HMSO, London, 1985–1986.
P. J. Hewitt, 'The BURL Guide to the Control of Substances Hazardous to Health Regulations 1988', H & H Scientific Consultants, Leeds, 1989.
'Human factors in industrial safety', HMSO, London.
A. W. Hunter, 'Restricted Medicines and Poisons', The Pharmaceutical Press, London, 1974.
R. P. Hunter, 'Diseases of Occupation', Hodder & Stoughton, London, 1978.
'IARC monographs on the evaluation of carcinogenic risk to humans' Supplement 7. World Health Organization, International Agency for Research in Cancer, 1987.

'Ink Fly in Newspaper Pressrooms', HMSO, London, 1984.

'Integrated Risk Information System', (U.S.) Environmental Protection Agency, EPA/600/6-86/032a.

'International Standards for Drinking Water', World Health Organization, Geneva, 1958.

'Introducing COSHH (a brief guide for all employers)', HMSO, London, n.d.

S. Kaye, 'Handbook of Emergency Toxicology', Charles C. Thomas, Springfield, Ill., 1977.

'Identification and Analysis of Organic Pollutants in Air', ed. L. Kieth, Butterworths, London, 1984.

'Kirk–Othmer Concise Encyclopaedia of Chemical Technology', John Wiley & Sons, New York, 1985.

'Methods for Biological Monitoring', ed. T. J. Kneip and J. V. Crable, American Public Health Association, Washington, D.C., 1988.

E. V. Kring *et al.*, 'A new passive colorimetric air-monitoring badge system for ammonia, sulphur dioxide and nitrogen oxide', *Am. Ind. Hyg. Ass. J.*, 1981, **42**, 373.

'Laboratory Biosafety Manual', World Health Organization, Geneva, 1983.

'Epidemiology: A Dictionary of Terms', ed. J. M. Last, International Epidemiological Association, c/o University of Ottawa, 1981.

'The Sigma-Aldrich Library of Chemical Safety Data, 2nd Edn', ed. R. E. Lenga, Sigma-Aldrich, Milwaukee, 1987.

D. Leon, 'Mortality of male members of two printing trade unions', Unpublished, 1989.

D. Leon, P. Thomas, and S. Hutchins, 'Lung cancer in newspaper machine men', Unpublished, 1989.

P. J. Lioy and M. J. Y. Lioy, 'Air Sampling Instruments', American Conference of Government Industrial Hygienists, Cincinnati, Ohio, 1989.

D. Mark, 'Problems associated with the use of membrane filters for dust sampling when compositional analysis is required', *Ann. Occup. Hyg.*, 1974, **17**, 35.

'Martindale: The Extra Pharmacopoeia', The Pharmaceutical Press, London, various dates.

'Proceedings of the United States Technical Conference on Air Pollution', ed. L. McCabe, McGraw-Hill, New York, 1952.

'McCrone Dispersion Staining Objective', McCrone Research Associates, London, n.d.

'Maximum Concentration at the Workplace and Biological Tolerance Values for Working Materials, 1989', VCH Publishers, Cambridge, 1989.

'Methods for Determination of Hazardous Substances', HMSO, London, various dates.

1 Acrylonitrile charcoal tube/gas chromatography (GC).

2 Acrylonitrile pumped thermal desorption/GC.

3 Generation of test atmospheres of organic vapours by the syringe injection technique.

4 Standard atmospheres permeation tube.

5 On-site validation of methods.

6 Lead atomic absorption (AA).

7 Lead *X*-ray fluorescence (XRF).

8 Lead colorimetric (dithizone).

9 Tetra alkyl lead personal monitoring.

10 Cadmium AA.
11 Cadmium XRF.
12 Chromium AA.
13 Chromium XRF.
14 Total inhalable and respirable dust gravimetric.
15 Carbon disulphide charcoal tube/GC.
16 Mercury adsorbent tube (Hydrar) AA.
17 Benzene charcoal tube/GC.
18 Tetra alkyl lead continuous monitoring.
19 Formaldehyde colorimetric (Chromotropic acid).
20 Styrene pumped charcoal tube/GC.
21 Glycol ethers charcoal tube/GC.
22 Benzene in air.
23 Glycol ethers thermal desorption/GC.
24 Vinyl chloride charcoal tube/GC.
25 Organic isocyanates reagent bubbler/HPLC.
26 Ethylene oxide charcoal tube/GC.
27 Diffusive sampler evaluation protocol.
28 Chlorinated hydrocarbon solvent vapours in air.
29 Beryllium AA.
30 Cobalt AA.
31 Styrene pumped thermal desorption/GC.
32 Phthalate esters solvent desorption/GC.
33 Adsorbent tube standards.
34 Arsine colorimetric (diethyldithiocarbamate).
35 HF and fluorides ion-selective electrode.
36 Toluene in air.
37 Quartz in respirable airborne dust direct infra-red.
38 Quartz in respirable airborne dust KBr disc technique.
39/2 Asbestos fibres light microscopy (European reference version).
39/3 Asbestos fibres in air.
40 Toluene in air.
41 Arsenic AA.
42 Nickel AA.
43 Styrene in air.
44 Styrene in air.
45 Ethylene dibromide solvent desorption/GC.
46 Platinum AA.
47 Rubber fume in air measured as total particulates and cyclohexane soluble material.
48 Newspaper print rooms: measurements of total particulates and cyclohexane soluble material in air.
49 Aromatic isocyanates acid hydrolysis-diazotization.
50 Benzene diffusive thermal desorption/GC.
51/2 Quartz in respirable dusts *X*-ray diffraction (direct method).
52/2 Hexavalent chromium in chromium plating mists colorimetric (1,5-diphenylcarbazide).
53 1,3-Butadiene thermal desorption/GC.
54 Protocol for assessing the performance of a pumped sampler for gases and vapours.

55 Acrylonitrile diffusive/thermal desorption/GC.
56/2 Hydrogen cyanide in air.
57 Acrylamide liquid chromatography.
58 Mercury vapour.
59 Manmade mineral fibres.
60 Mixed hydrocarbons.
61 Total hexavalent chromium compounds in air colorimetric.
62 Aromatic carboxylic acid anhydrides.
63 Butadiene Molecular Sieve/thermal desorption/GC.
64 Toluene charcoal diffusion/solvent desorption/GC.
65 Mine road dust: determination of incombustible matter.
66 Mixed hydrocarbons (C_5 to C_{10}) in air thermal desorption/GC.
67 Total (and speciated) chromium in chromium plating mists colorimetric (1,5-diphenylcarbazide).
68 Coal tar pitch volatiles.
69 Toluene in air.

'Methods of Air Sampling and Analysis', American Public Health Association Intersociety Committee, Washington, D.C., 1972.

'Asbestos. Vol. I Properties, Application and Hazards', ed. L. Michaels and S. S. Chissick, John Wiley & Sons, Chichester, 1979.

R. Mitchell, 'Introduction to Environmental Microbiology', Prentice Hall, New Jersey, 1974.

D. S. Murray, 'Man's Microbic Enemies', Watts, London, 1932.

'Developments in Industrial Microbiology. Proceedings of the 8th General Meeting of the Society for Industrial Microbiology, 1971', ed. E. D. Murray, American Institute of Biological Sciences, Washington, D.C., 1972.

'Proceedings of the 9th General Meeting of the Society for Industrial Microbiology, 1972', American Institute of Biological Sciences, Washington, D.C., 1973.

'Information Leaflet No. 9', National Pharmaceutical Association, St. Albans, n.d.

Newsletter, Summer 1989, p 7, The Association of Consulting Scientists, London.

'NIOSH Handbook of Statistical Tests for Evaluating Employee Exposure to Air Contaminants', HEW Publication No. 75-147, U.S. Department of Health and Human Services, Washington, D.C., 1975.

'NIOSH Manual of Analytical Methods', U.S. Department of Health and Human Services, Washington, D.C., 1984.

'NIOSH Pocket Guide to Chemical Hazards', U.S. Department of Health and Human Services, Washington, D.C., 1987.

'Official Methods of Analysis', 14th Edn. Association of Official Analytical Chemists, Washington, D.C., 1984.

'Proceedings of the 1959 Pneumoconiosis Conference, Johannesburg', ed. A. J. Orensten, Churchill, London, 1960.

K. Pannwitz, 'Direct reading diffusion tubes', *Drager Review* 1984, **53**, 10.

F. A. Patty, 'Industrial Hygiene and Toxicology', Wiley Interscience, New York, 1981.

D. J. Paustenbach, ed., 'The Risk Assessment of Environmental Hazards', John Wiley & Sons, New York, 1989.

'Chemistry for Protection of the Environment', ed., L. Pawlowski, A. J. Verdier and W. J. Lacy, Elsevier, Amsterdam, 1984.

'Poisons Act', 1972, and 'Poisons Rules and Orders', 1982, HMSO, London.

'Hazardous Waste Management Handbook', ed. A. Porteus, Butterworths, London, 1985.

N. H. Proctor and J. P. Hughes, 'Chemical Hazards of the Workplace', Lippincott, Philadelphia, 1978.

'Professional Conduct: Guidance for Chemists', The Royal Institute of Chemistry, London, 1975.

'Prudent Practices for Handling Hazardous Chemicals in Laboratories', National Research Council, National Academy Press, Washington, D.C., 1981.

A. R. Purdy and D. H. Williams, 'Fibre Identification Using Optical Microscopy', The Electricity Council (Safety Review Supplement), London, 1977.

'Community Documentation Centre on Industrial Risk, Vol. 1', ed. K. Rasmussen, Commission of the European Communities, Ispra, April 1989.

'Community Documentation Centre on Industrial Risk, Vol. 2', ed. K. Rasmussen, Commission of the European Communities, Ispra, August 1989.

'Recombinant DNA Safety Considerations', Organization for Economic Co-operation and Development, Paris, 1986.

'Record Keeping Book for COSHH', Croner Publications, New Malden, 1989.

'Reports on Public Health and Medical Subjects No. 71. The Bacteriological Examination of Water Supplies', 198 (revised and enlarged) HMSO, London, 1969.

J. Reynolds, 'Pilot morbidity study of Fleet Street production workers', 1988.

'Martindale: The Extra Pharmacopoeia', 29th Edn., ed. J. E. F. Reynolds, The Pharmaceutical Press, London, 1989.

M. L. Richardson, 'Risk Assessment for Hazardous Installations', Pergamon, Oxford, 1986.

'Toxic Hazard Assessment of Chemicals', ed. M. L. Richardson, The Royal Society of Chemistry, London, 1986.

'Risk Assessment of Chemicals in the Environment', ed. M. L. Richardson, The Royal Society of Chemistry, London, 1988.

'Chemistry, Agriculture and the Environment', ed. M. L. Richardson, The Royal Society of Chemistry, Cambridge, 1991.

'IUPAC—Glossary of Terms in Chemical Toxicology and Ecotoxicity', ed. M. L. Richardson and J. H. Duffus, in press.

A. H. Rose, 'Chemical Microbiology', Butterworths, London, 1968.

N. I. Sax and R. J. Lewis Senior, 'Dangerous Properties of Industrial Materials', 7th Edn., Van Nostrand Reinhold, New York, 1988.

D. Simpson and W. G. Simpson, 'An Introduction to Applications of Light Microscopy in Analysis', The Royal Society of Chemistry, London, 1988.

'Croner's Health and Safety at Work', ed. P. Smith, Croner Publications, New Malden, revised regularly.

'Safety Lines', ed. P. Smith and P. Reeve, Croner Publications, New Malden, various dates.

'Browning: Toxicity and Metabolism of Industrial Solvents Vol. I: Hydrocarbons', ed. R. Snyder, Elsevier, Amsterdam, 1987.

'Standard Methods for the Examination of Water and Waste Water', American Public Health Association, Washington, D.C., 1976.

'Statutory Instruments', HMSO, London, dates shown.

No. 37, 1987. The Dangerous Substances in Harbour Regulations.

No. 976, 1985. Social Security (Industrial Injuries) (Prescribed Diseases) Regulations, and Subsequent Amendment Regulations 1986 and 1987.

No. 1059, 1981. The Dangerous Substances (Conveyance by Road Tankers and Tank Containers) Regulations.

No. 1244, 1984. Health and Safety, The Classification, Packaging and Labelling of Dangerous Substances Regulations.
No. 1248, 1980. Health and Safety, The Control of Lead at Work Regulations.
No. 1657, 1988. Health and Safety, The Control of Substances Hazardous to Health Regulations.
No. 1709, 1980. Public Health, England Wales, and Scotland, The Control of Pollution (Special Waste) Regulations.
No. 1810, 1989. Health and Safety (Genetic Manipulation) Regulations.
No. 1951, 1986. The Road Traffic (Carriage of Dangerous Substances in Packages, *etc.*) Regulations.

F. B. Stern, L. I. Murthy, J. J. Beaumont, P. A. Schulte, and W. E. Halperin, 'Notification and Risk Assessment for Bladder Cancer of a Cohort Exposed to Aromatic Amines', *J. Occup. Med.*, 1985, **27**, pp 495–500.

'Substances for use at work: the provision of information', HMSO, London, n.d.

'Survey of Compounds which have been Tested for Carcinogenic Activity', U.S. Department of Health, Washington, D.C., annual volumes.

'Tactile directive meets mixed industry response', *Packaging Week*, 1990, **36**, pp 1–3.

'Technical Note 1: Measurement of airborne asbestos dust by the membrane filter method', Asbestosis Research Council, London, 1978.

'Technical Note 3: Recommendations for the sampling and identification of asbestos in asbestos products', Asbestosis Research Council, London, 1978.

'The PHIPCO Manual', The Public Health and Industrial Pesticides Council, Sheringham, 1989.

'The provision of health and safety information by manufacturers, importers and suppliers of chemical products to the printing industry', HMSO, London, 1986.

'The Retail Pocket Book', NTC Publications, Henley-on-Thames, 1991.

'The Selection and Use of Personal Sampling Pumps', Technical Guide No. 5, British Occupational Hygiene Society, Northwood: Science Reviews, 1985.

The Spectator, London, 22.9.90, p 32.

'Geochemistry and Health', ed. I. Thornton, Northwood: Science Reviews, 1988.

'Threshold Limit Values and Biological Exposure Indices 1989–1990', American Conference of Government Industrial Hygienists, Cincinnati, Ohio, 1989.

R. A. Trevethick, 'Environmental and Industrial Health Hazards—a Practical Guide', Heinemann Medical, London, 1976.

'(U.S.) Manufacturing Chemists' Association, Safety & Fire Protection Committee, Guide for Safety in the Chemical Laboratory', Van Nostrand Reinhold, New York, 1972.

'(U.S.) National Fire Protection Association Guide to Hazardous Materials'.

U.S. Public Health Service Centers for Disease Control/National Institutes of Health (see: Biosafety in Microbiological and Biomedical Laboratories).

N. P. Vaughan *et al.*, 'Filter weighing reproducibility and the Gravimetric Detection Limit', *Ann. Occup. Hyg.*, 1989, **33**, 3313.

R. L. Vollum *et al.*, 'Fairbrother's Textbook of Bacteriology', Heinemann, London, 1970.

'CRC Handbook of Chemistry and Physics', ed. R. C. Weast, CRC Press, Boca Raton, 1989.

Welsh Office (see: Reports on Public Health).

B. H. Woolen *et al.*, 'Human inhalation pharmacokinetics of 1,1,2-trichloro 1,2,2-trifluoroethane (Fluorocarbon 113)', *Int. Arch. Occup. Environ. Health*, 1990, **62**, 73.

O. Wong, R. W. Morgan, W. J. Bailey *et al.*, 'An Epidemiological Study of Petroleum Industry Employees', *Br. J. Ind. Med.*, 1986, **43**, 6–17.

'Workplace Analysis Scheme for Proficiency—Information for Participants', Health and Safety Executive Committee on Analytical Requirements, 1988. Occupational Medicine & Hygiene Laboratory, 403/405 Edgware Road, London, NW2 6LN.

World Health Organization (see: IARC monographs, International Standard for Drinking Water, Laboratory Safety Manual, *etc.*).

'The Pesticide Manual A World Compendium, 8th Edn.', C. R. Worthing and S. B. Walker, BCPC Publications, Berkshire, 1987.

Subject Index